Probability, Grade 7
A complete workbook with lessons and problems

By Maria Miller

Copyright 2017-2020 Maria Miller.
ISBN 978-1532945601

EDITION 11/2020

All rights reserved. No part of this workbook may be reproduced or transmitted in any form or by any means, electronic or mechanical, or by any information storage and retrieval system, without permission in writing from the author.

Copying permission: Permission IS granted for the teacher to reproduce this material to be used with students, not for commercial resale, by virtue of the purchase of this workbook. In other words, the teacher MAY make copies of the pages to be used with students.

Contents

Preface	5
Introduction	7
Helpful Resources on the Internet	8
Probability	13
Probability Problems from Statistics	16
Experimental Probability	18
Counting the Possibilities	21
Using Simulations to Find Probabilities	27
Probabilities of Compound Events	33
Review	37
Answers	39
Appendix: Common Core Alignment	53

Preface

Hello! I am Maria Miller, the author of this math book. I love math, and I also love teaching. I hope that I can help you to love math also!

I was born in Finland, where I also grew up and received all of my education, including a Master's degree in mathematics. After I left Finland, I started tutoring some home-schooled children in mathematics. That was what sparked me to start writing math books in 2002, and I have kept on going ever since.

In my spare time, I enjoy swimming, bicycling, playing the piano, reading, and helping out with Inspire4.com website. You can learn more about me and about my other books at the website MathMammoth.com.

This book, along with all of my books, focuses on the conceptual side of math... also called the "why" of math. It is a part of a series of workbooks that covers all math concepts and topics for grades 1-7. Each book contains both instruction and exercises, so is actually better termed *worktext* (a textbook and workbook combined).

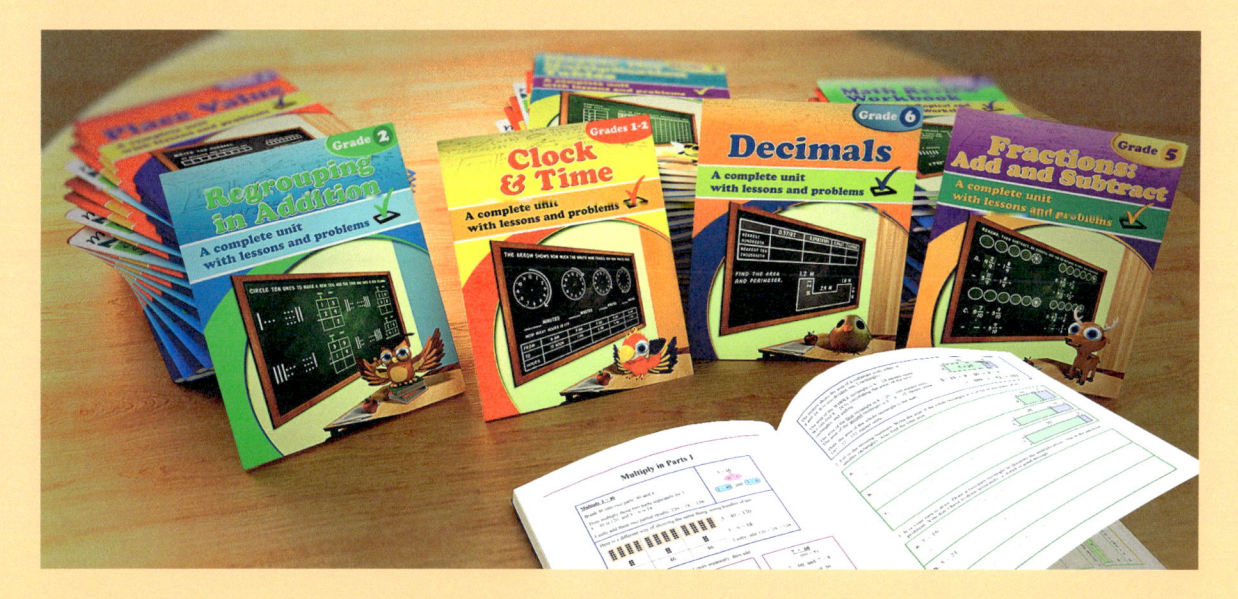

My lower level books (approximately grades 1-5) explain a lot of mental math strategies, which help build number sense — proven in studies to predict a student's further success in algebra.

All of the books employ visual models and exercises based on visual models, which, again, help you comprehend the "why" of math. The "how" of math, or procedures and algorithms, are not forgotten either. In these books, you will find plenty of varying exercises which will help you look at the ideas of math from several different angles.

I hope you will enjoy learning math with me!

Introduction

The concepts covered in *Probability, Grade 7 Workbook* are very likely new to your students. However, most students have an intuitive understanding of probability based on hearing the terms "probably" and "likely," listening to weather forecasts, and so on.

In the past, probability wasn't taught until high school—for example, I personally encountered it for the first time in 12th grade. However, since probability is such a useful and easily accessible field of math, it was felt that it should be introduced sooner, so during the 1990s and 2000s it "crept" down the grade levels until many states required probability even in elementary school. The Common Core Standards include probability starting in 7th grade. I feel that is good timing because by 7th grade students have studied fractions, ratios, and proportions, so they have the tools they need to study probability. Moreover, they will need an understanding of the basic concepts of probability in order to understand the statistical concepts that they will study in middle school and high school.

In this workbook, we start with the concept of simple (classic) probability, which is defined as the ratio of the number of favorable outcomes to the number of all possible outcomes. Students calculate probabilities that involve common experiments, which include flipping a coin, tossing a pair of dice, picking marbles, and spinning a spinner.

The lesson *Probability Problems from Statistics* introduces probability questions involving the phrase "at least," which are often solved by finding the probability of the complement event. For example, it might be easier to count the number of students who got at most D+ on a test than to count the number of students who got at least C-.

In the next lesson, *Experimental Probability*, students conduct experiments, record the outcomes, and calculate both the theoretical and experimental probabilities of events, in order to compare the two. They will draw a card from a deck or roll a die hundreds of times. You can access simulations of these activities at the web page
https://www.mathmammoth.com/lessons/probability_simulations.php

Next, we study compound events, which combine two or more individual simple events. Tossing a die twice or choosing first a girl then a boy from a group of people are compound events. Students calculate the probabilities of compound events by using the complete sample space (a list of all possible outcomes). They construct the sample space in several ways: by drawing a tree diagram, by making a table, or simply by using logical thinking to list all the possible outcomes.

The last major topic in this workbook is simulations. Students design simulations to find the probabilities of events. For example, we let heads represent "female" and tails represent "male," so we can toss a coin to simulate the probability of choosing a person of either sex at random. Later in the lesson, students design simulations that use random numbers. They generate those numbers by using either the free tool at https://www.random.org/integers or a spreadsheet program on a computer.

In the last lesson of the workbook, *Probabilities of Compound Events*, we learn to calculate the probability of a compound event by *multiplying* the probabilities of the individual events (assuming the outcomes of the individual events are independent of each other). This topic exceeds the Common Core Standards for 7th grade and thus is optional. I have included it here because the idea studied in the lesson is very simple and I feel many students will enjoy it.

I wish you success in teaching math!

Maria Miller, the author

Helpful Resources on the Internet

Use these free online games and resources to supplement the "bookwork" as you see fit.

Probability Videos by Maria
These video lessons cover topics that have been chosen to complement the lessons in this workbook.
https://www.mathmammoth.com/videos/probability/probability_lessons.php

SIMPLE PROBABILITY

Probability Game with Coco
A multiple-choice online quiz on simple probability.
https://www.math-play.com/Probability-Game.html

Probability Circus
Choose the spinner that matches the probability in this interactive online activity.
http://www.hbschool.com/activity/probability_circus/

Simple Probability
Practice finding probabilities of events, such as rolling dice, drawing marbles out of a bag, and spinning spinners.
https://www.khanacademy.org/math/cc-seventh-grade-math/cc-7th-probability-statistics/cc-7th-basic-prob/e/probability_1

Card Sharks Game
Use your knowledge of probability to bet on whether the next card is higher or lower than the last one.
https://mrnussbaum.com/card-sharks-online-game

Mystery Spinners
In this activity, you need to find the hidden mystery spinner working from only one clue.
https://www.scootle.edu.au/ec/viewing/L2384/index.html

Simple Probability Quiz
Reinforce your probability skills with this interactive self-check quiz.
http://www.phschool.com/webcodes10/index.cfm?wcprefix=ara&wcsuffix=1201

EXPERIMENTAL PROBABILITY

Probability Tools
Play around with dice, coins, spinners, playing cards, counters and digit cards. You can choose from a variety of settings such as number of dice, number of trials, whether to display results as a frequency diagram or table.
https://www.interactive-maths.com/probability-tools-flash.html

Adjustable Spinner
Create a virtual spinner with variable-sized sectors to compare experimental results to theoretical probabilities. You can choose the sizes of the sectors, the number of sectors, and the number of trials.
http://www.shodor.org/interactivate/activities/AdjustableSpinner/

Interactive Customizable Spinners
Use a tool to build colored spinners. Then, test the spinner over a number of spins. Compare the actual results with the expected results.
https://fuse.education.vic.gov.au/Resource/LandingPage?ObjectId=8eb446b6-bd1e-446a-848a-935dce8b0b70

Dice Duels Tool
Explore the experimental probability distributions when you roll between two and five dice, and either add, subtract, or multiply the numbers. The tool graphs the results, and can do up to 9,999 rolls.
https://fuse.education.vic.gov.au/Resource/LandingPage?ObjectId=f8341288-4733-4604-abf3-7c1d9de7fc4b

Dice Roll
Choose the number of virtual dice to roll and how many times to roll them. The page shows both the actual results and expected (theoretical) probabilities. The simulation works for a very large numbers of rolls.
http://www.btwaters.com/probab/dice/dicemain3D.html

Coin Flip
This virtual coin toss shows the results numerically and can generate at least 100,000 flips.
http://www.btwaters.com/probab/flip/coinmainD.html

Coin Toss Simulation
Another virtual coin toss. This one shows the results both using images of coins and numerically.
https://syzygy.virtualave.net/multicointoss.htm

Rocket Launch
All three stages of the rocket launch must successfully pass their pre-takeoff tests. By default, each stage has a 50% chance of success, however, this can be altered by dragging the bar next to each stage. Observe how many tries it takes until there is a successful launch.
https://mste.illinois.edu/activity/rocket/

Find the Bias
A die (one cube of dice) has been weighted (loaded) to favor one of the six numbers. Roll the die to work out which is the favored face. Explore how many rolls are needed for you to be reasonably sure of a conclusion.
https://fuse.education.vic.gov.au/Resource/LandingPage?ObjectId=0ec1fe96-7c91-4975-9863-766a1fe9c1c5

Load One Dice (Biased dice)
Make a biased die by loading it to favor one of the six numbers. Then roll the loaded die, and compare the shape of theoretical data distributions with experimental results.
https://fuse.education.vic.gov.au/Resource/LandingPage?ObjectId=b60d5135-608d-4fd6-8e65-a81fa1dc172a

Racing Game with One Die
Explore how experimental probability relates to fair and unfair games. You choose which and how many numbers of the die make each of the cars move. Other options include the number of trials and the length of the race. The program calculates the percentage of wins for each car and draws a pie chart.
http://www.shodor.org/interactivate/activities/RacingGameWithOneDie/

COMPOUND EVENTS AND TREE DIAGRAMS

Tree Diagrams and Spinners Quiz
Practice reading tree diagrams in this interactive 10-question quiz.
https://www.thatquiz.org/tq/practicetest?8x24p16y3o5t

Lucky 16 Game
You place counters on the game board, and then they will be removed based on the sum of two dice that are rolled. Try to predict the best positions for the counters before the game starts.
https://fuse.education.vic.gov.au/Resource/LandingPage?ObjectId=31ee684d-742e-4fe3-bb55-e8e6e61e0d6d

Airport Subtraction Game
This game is based on rolling two dice and subtracting the results. You task is to place your plane at the back of the queue on one of the runways. Try to predict which lane is most likely to clear quickly.
https://fuse.education.vic.gov.au/Resource/LandingPage?ObjectId=0c1b063f-52ec-4564-ac6e-265ddcbdcada

Quiz: Compound Probability
Test your knowledge of compound probability with this interactive self-check quiz.
http://www.phschool.com/webcodes10/index.cfm?wcprefix=aba&wcsuffix=1205

How could I send the check and not pay the bill?
What is the probability that Tessellation will put each of the three checks into the correct envelopes if she does it randomly? The page includes a hint and a complete solution (click "answer" at the bottom of page).
https://figurethis.nctm.org/challenges/c69/challenge.htm

Flippin' Discs – interactive activity
You throw two discs. You win if they both show the same color. You can run the game 100 times and see the detailed results. Can you explain why you win approximately half the time? Explore the situation also with 3, 4, and even 5 discs. The solution is found in a link near the top left of the page.
https://nrich.maths.org/4304

"Data Analysis & Probability Games" from MathWire
A list of board and dice games to help to teach topics appropriate for beginners in probability.
http://mathwire.com/games/datagames.html

Cross the Bridge
This is a printable board game based on throwing two dice and the probabilities for the sum of the dice.
http://www.mathsphere.co.uk/downloads/board-games/board-game-17-crossing-the-river.pdf

At Least One…
The tree diagram and related discussion on this page helps students answer probability questions like, "What is the probability of getting at least one head by flipping a coin ten times?" A link to the solution is near the top left of the page.
https://nrich.maths.org/7286

SIMULATIONS

Probability Simulations in Excel
These spreadsheet files match some of the lengthier probability simulations in this workbook.
https://www.mathmammoth.com/lessons/probability_simulations.php

Random Integer Generator
Choose how many numbers, how many columns, and the values of the integers and then click to generate.
https://www.random.org/integers/

Coin Toss Simulation
Another virtual coin toss. This one shows the results both using images of coins and numerically.
https://syzygy.virtualave.net/multicointoss.htm

Marbles
Run repeated experiments where you draw 1, 2, or 3 marbles from a set of blue, red, purple, and green marbles. The results shown include the frequencies of each possible outcome, the experimental probabilities, and the theoretical probabilities. This activity can be used not only to explore probabilities but also to perform simulations.
http://www.shodor.org/interactivate/activities/Marbles/

Interactivate: *Fire!!* and *Directable Fire!!*
In these two activities, you first set the probability that a fire will spread from tree to tree in a forest of 100 trees. Then you click the tree where the fire starts and watch it spread. In the Directable Fire activity, you can set the probabilities for each direction to be different. Repeat the activity several times to see that the amount of forest that burns varies (for any set probability of fire spreading).
http://www.shodor.org/interactivate/activities/Fire/

http://www.shodor.org/interactivate/activities/DirectableFire/

COMPOUND PROBABILITY

Probability Quiz
Test your knowledge about probability in this interactive self-check quiz.
http://www.phschool.com/webcodes10/index.cfm?wcprefix=ama&wcsuffix=7454

Probability of Compound Events Quiz
Practice finding the probability of compound events. Some problems deal with replacement and non-replacement.
https://maisonetmath.com/probability/quizzes/409-probability-of-compound-events

Probability Quiz
Reinforce your skills with this interactive multiple-choice quiz.
http://www.phschool.com/webcodes10/index.cfm?wcprefix=aqa&wcsuffix=1005&area=view

FOR FUN

Monty Hall Paradox
Try this interactive version of the famous Monty Hall problem. Behind which door is the car? If you choose the wrong one, you'll win a goat instead.
https://www.math.ucsd.edu/~crypto/Monty/monty.html

What Does Random Look Like?
This problem challenges our thinking about randomness. Make up a sequence of twenty Hs and Ts that *could* represent a sequence of heads and tails generated by a fair coin. Then use the animation to generate truly random sequences of 20 coin flips. Can you learn how to spot fakes?
https://nrich.maths.org/7250

Same Number
Imagine you are in a class of thirty students. The teacher asks everyone to secretly write down a whole number between 1 and 225. How likely is it for everyone's numbers to be different? The web page provides an interactive simulation so you can experiment with this problem. The following discussion also leads students to the classic birthday problem. The solution is found in a link near the top left of the page.
https://nrich.maths.org/7221

Probability

You *probably* already have an intuitive idea of what *probability* is. In this lesson we look at some simple examples in order to study probability from a mathematical point of view.

If we flip a coin, the chance, or **probability**, of getting "heads" is 1/2, and the chance of getting "tails" is also 1/2. "Heads" and "tails" are the two possible **outcomes** when you toss a coin, and they are equally likely.

When rolling a six-sided number cube (a die), you have six possible **outcomes**: you can roll either 1, 2, 3, 4, 5, or 6. These are all equally likely (assuming the die is fair).

Thus the probability of rolling a five is 1/6. The probability of rolling a three is also 1/6. In fact, the probability of each of the six outcomes is 1/6.

The probability of rolling an even number is 3/6, or 1/2, because three of the six possible outcomes are even numbers.

Simple probability has to do with situations where each possible outcome is <u>equally likely</u>.

Then the **probability** of an event is the fraction $\dfrac{\text{number of favorable outcomes}}{\text{number of possible outcomes}}$

"Favorable outcomes" are those that make up the event you want. The examples will make this clear.

Example 1. What is the probability of getting a number that is less than 6 when tossing a fair die?

Count how many of the outcomes are "favorable" (less than 6). There are five: 1, 2, 3, 4, or 5.
And there are six possible outcomes in total.

Therefore, the probability is $\dfrac{\text{number of favorable outcomes}}{\text{number of possible outcomes}} = \dfrac{5}{6}$.

In math notation we write "P" for probability and put the event in parentheses:

P(less than 6) = 5/6.

Example 2. On this spinner the number of possible outcomes is eight, because the arrow is equally likely to land on any of the eight wedges. What is the probability of spinning yellow?

There are TWO favorable outcomes (yellow areas) out of EIGHT possible outcomes (wedges).

P(yellow) = 2/8 = 1/4.

(Because green and yellow each have two wedges, there are only six possible colors that can result.
When we list the possible outcomes, we list the six colors. But when we figure the probabilities, we must use the eight equally-probable wedges.)

By convention, the probability of an event is always at least 0 and at most 1. In symbols: $0 \leq P(\text{event}) \leq 1$.

A probability of 0 means that the event does not occur; it is impossible. Probability of 1 means that the event is sure to occur; it is certain. A probability near 1 (such as 0.85) means that the event is likely to occur. A probability of 1/2 means that an event is neither likely nor unlikely.

Example 3. What is the probability of rolling 8 on a standard six-sided die?

This is an impossible event, so its probability is zero: P(8) = 0.

Example 4. What is the probability of rolling a whole number on a die?

This is a sure event, so its probability is one. P(whole number) = 1.

1. There are three red marbles, two dark blue marbles, and five light green marbles in Michelle's bag. List all the possible outcomes if you choose one marble randomly from her bag.

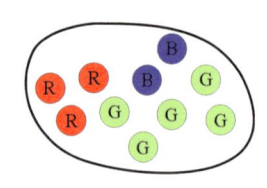

2. Michelle chooses one marble at random from her bag. What is the probability that...

    a. the marble is blue?

    b. the marble is not red?

    c. the marble is neither blue nor green?

3. Make up an event with a probability of zero in this situation.

4. Suppose you choose one letter randomly from the word "PROBABILITY."

    a. List all the possible outcomes for this event.

    Now find the probabilities of these events:

    b. P(B)

    c. P(A or I)

    d. P(vowel)

    e. Make up an event for this situation that is likely to occur, yet not a sure event, and calculate its probability.

> **The complement of an event and the probability of "not"**
>
> The **complement** of any event A is the event that A does *not* occur.
>
> If the probability of event A is a, then the probability of A not happening is simply $1 - a$.

5. The weatherman says that the chance of rain for tomorrow is 1/10. What is the probability of it not raining?

6. The spinner is spun once. Find the probabilities as simplified fractions.

    a. P(green)　　　　　　　　**b.** P(not green)

    c. P(not pink)　　　　　　　**d.** P(not black)

    e. Make up an event for this situation that is not likely, yet not impossible either, and calculate its probability.

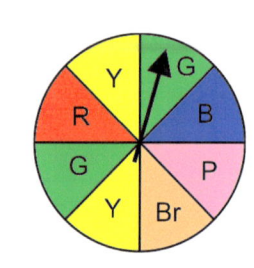

Probabilities are often given as percentages instead of fractions.

Example 5. Kimberly's sock bin contains 7 brown socks, 9 white socks, and 5 red socks. She picks one without looking. What is the probability that she gets a white sock?

There are 9 white socks out of 21 socks in all. The probability is $9/21 = 3/7 = 3 \div 7 \approx 0.42857 = 0.429 = 42.9\%$.

7. Suppose you were to draw one card from the set of cards on the right. Complete the table with the possible outcomes, and their probabilities both as fractions and as percentages (to the nearest tenth of a percent).

Possible outcomes	Probability (fraction)	Probability (percentage)

8. This "rainbow spinner" is spun once. Find the probabilities to the nearest tenth of a percent.

   a. P(yellow)

   b. P(blue or green)

   c. P(not orange)

   d. P(not red and not purple)

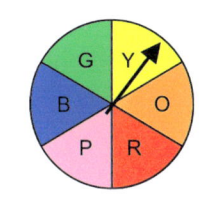

   e. Make up an event for this situation with a probability of 1.

9. **a.** An empty bus has 45 seats, and 22 of them are window seats. If you are assigned a seat at random, what is the probability, to the nearest tenth of a percent, that you get a window seat?

   b. Now each seat marked with an "x" is already occupied. If you choose a seat randomly, what is the probability, to the nearest tenth of a percent, that you get a window seat?

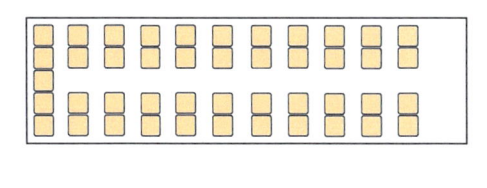

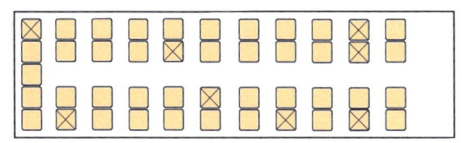

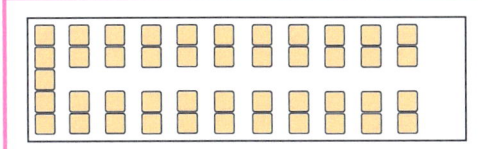

The chart shows you the seating arrangement of a bus. You enter the bus, and the driver informs you that fifteen seats are already occupied and that if you choose a seat randomly, the probability of getting a window seat is less than 25%.

How many window seats are occupied, at least?

Puzzle Corner

Probability Problems from Statistics

Example 1. The bar graph shows the science test scores of all seventy 7th graders in Westmont School. If you choose one of them at random, then what is the probability that the student's score was at least C− (in other words, C− or better)?

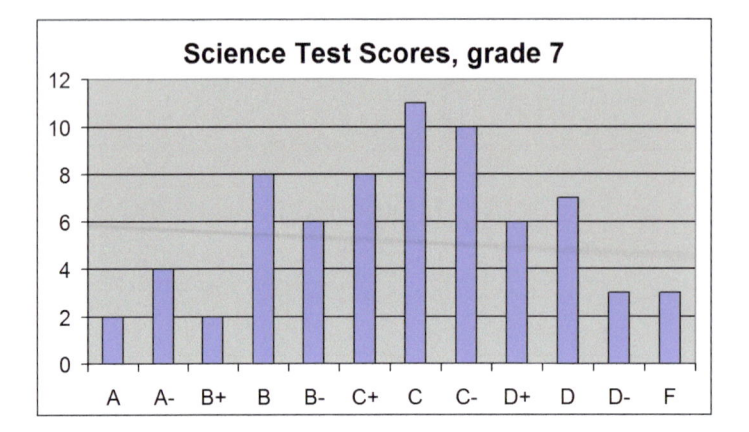

Sometimes when a probability question involves "at least," it is easier to look at the complement event — everything else — and find its probability first. The complement of "at least C−" is "at most D+" in other words, D+, D, D−, and F. From the graph, it is easier to sum the number of students who got the four low scores than to sum the number of students who got the eight high scores.

The number of students who got D+, D, D−, or F is 6 + 7 + 3 + 3 = 19 students. There are a total of 70 students, so P(at most D+) = 19/70. Now it's easy to calculate the original probability in question: P(at least C−) = 1 − 19/70 = 51/70.

1. You choose one student at random from the 7th graders in Westmont School.

   a. What is the probability that the student's score was at least D?

   b. What is the probability that the student's score was at most B+?

2. The dotplot shows the age distribution of a children's fishing club. One child is chosen randomly from the group.

   a. What is the probability that the child is at most 9 years of age?

   b. What is the probability that the child is at least 7 years of age?

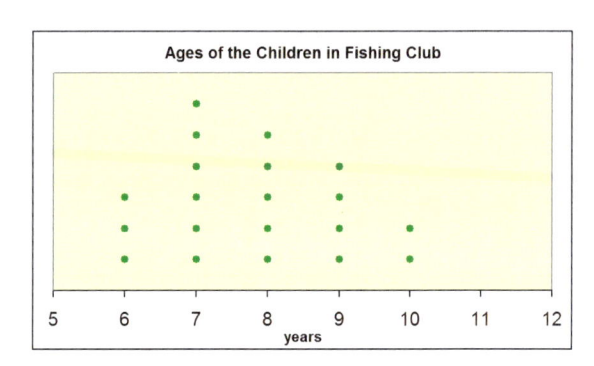

3. The chart lists the languages most commonly spoken at home in the United States and the number of people at least 5 years old who speak them.

   If one person is selected randomly from this population, what is the probability, to the nearest hundredth of a percent, that the person does *not* speak only English at home?

Only English	215,423,557
Spanish	28,101,052
Other Indo-European	10,017,989
Asian Language	6,960,065
Other	1,872,489
Total Population Age 5+	262,375,152

4. The table lists the numbers of males and females in various age groups in the United States in the year 2000. Answer the questions to the nearest tenth of a percent.

   a. What is the probability that a randomly chosen *person* in the United States is a female, 25-29 years old?

   b. Now look at the females only. What is the probability that a randomly chosen US female is 25-29 years old?

   c. What is the probability that a randomly chosen person in the United States is a male who is at least 15 years old?

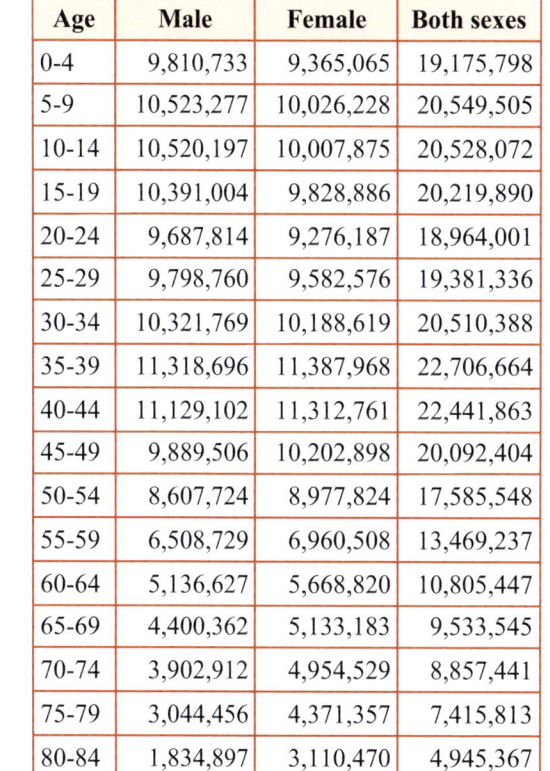

   d. What is the probability that a randomly chosen person in the United States is at most 64 years old?

Age	Male	Female	Both sexes
0-4	9,810,733	9,365,065	19,175,798
5-9	10,523,277	10,026,228	20,549,505
10-14	10,520,197	10,007,875	20,528,072
15-19	10,391,004	9,828,886	20,219,890
20-24	9,687,814	9,276,187	18,964,001
25-29	9,798,760	9,582,576	19,381,336
30-34	10,321,769	10,188,619	20,510,388
35-39	11,318,696	11,387,968	22,706,664
40-44	11,129,102	11,312,761	22,441,863
45-49	9,889,506	10,202,898	20,092,404
50-54	8,607,724	8,977,824	17,585,548
55-59	6,508,729	6,960,508	13,469,237
60-64	5,136,627	5,668,820	10,805,447
65-69	4,400,362	5,133,183	9,533,545
70-74	3,902,912	4,954,529	8,857,441
75-79	3,044,456	4,371,357	7,415,813
80-84	1,834,897	3,110,470	4,945,367
85+	1,226,998	3,012,589	4,239,587
Totals	138,053,563	143,368,343	281,421,906

Experimental Probability

> In this lesson, we study **experimental probability**, which refers to the probability of an event based on actually conducting an experiment and observing how often the event occurs. It is the ratio of the number of times an event occurred to the number of times tested.
>
> This is in contrast to **theoretical probability**, which is calculated using mathematics and logical thinking, without actually conducting any experiments.

1. In this exercise, you will roll a die a lot of times to explore whether the chance of getting 1, 2, 3, 4, 5, and 6 is indeed 1/6 like it is theoretically.

    a. You will be rolling a (fair) die 60 times. Based on the theoretical probabilities, we would expect that each of the numbers 1, 2, 3, 4, 5, and 6 would come up exactly 1/6 · 60 = 10 times. Will that happen in reality?

    Roll a die 60 times and record each outcome. Count the **frequency** (how many times) that each number occurred and list the results in the table below. In the last column, calculate the percentages of how often each outcome occurred, to the hundredth of a percent.

    b. Now you will be rolling a die 120 times. How many times would you expect to roll each number, based on the theoretical probabilities of 1/6? _____ times

    However, we know that that will probably not happen. We can predict that each number will be rolled roughly that many times, but probably not exactly that many.

    Roll a die 120 times (start your count anew), record the outcomes again, and fill in the table for part (b). (Use multiple dice and several persons rolling them for quicker results.)

    c. One more time: roll the die 480 times, record the outcomes, and fill in the table for part (c).

a. With 60 rolls:			b. With 120 rolls:			c. With 480 rolls:		
Outcome	Frequency	Probability	Outcome	Frequency	Probability	Outcome	Frequency	Probability
1			1			1		
2			2			2		
3			3			3		
4			4			4		
5			5			5		
6			6			6		
TOTALS	60	100%	TOTALS	120	100%	TOTALS	480	100%

   d. Theoretically, the probability of rolling any of the numbers is 1/6 or 16.67%. In which of the three experiments — rolling the die 60 times, 120 times, or 480 times — were the experimental probabilities closest to the theoretical?

2. Through the marvels of automation, you will now "roll" a dice more times than in Exercise 1. You can use a spreadsheet file (#1) from the list at

https://www.mathmammoth.com/lessons/probability_simulations.php

Or you can use this virtual dice roller:
http://www.btwaters.com/probab/dice/dicemain3D.html

Note: If you roll two or more dice in this simulation, the results show the sum *of the dots on the dice, not the actual numbers that were rolled. So you'll want to leave the input value for "number of dice" set to "1."*

a. Predict about how many times you expect to get each of the six possible numbers if you roll a die 1,000 times:

About _____ times

b. Now roll one die 1,000 times.

If you use the virtual roller, choose "session" (and not "historical") to see the data for your session. To rerun the simulation you need to refresh the page (press F5).

Record in the table the frequencies of each outcome and calculate experimental probabilities. Observe how close each experimental probability is to the theoretical probability of 1/6 = 16.67%.

c. Let's say you were to roll a die 5,000 times. How do you expect the results to differ from rolling a dice 1,000 times?

Outcome	Frequency	Experimental Probability (%)
1		
2		
3		
4		
5		
6		
TOTALS	1,000	100%

d. (Optional) Try some larger simulations to see if you can determine about how many rolls it takes for the experimental probabilities to fall within half a percentage point of the theoretical value (from 16.17% to 17.17%)?

3. Select a set of 12 cards from a deck of playing cards so that the set has five different kinds of cards in it. Choose numbers from different suits. For example, you could choose the cards 2, 2, 2, 3, 3, 4, 5, 5, 5, 5, 6, and 6. The experiment will involve choosing a card at random from your set.

a. Choose a card at random from your set, record the outcome, and put the card back. Repeat this 100 times. Count the frequencies of each outcome. In the table, record the **relative frequencies**—the frequencies written as fractions of the total number of repetitions.

Outcome	Theoretical probability	Relative Frequency	Experimental probability
	_____%	_____/100	_____%

b. Calculate the theoretical and experimental probabilities for each outcome. Then compare the two: are they fairly close?

If not, what could have caused the discrepancy?

4. You will now conduct an experiment where the various outcomes are not equally likely to occur. In such a case, we say that the probability model is **not uniform**. Choose from one of the experiments listed or come up with one of your own. Repeat the experiment 100 times and count how many times each outcome occurs. Then calculate the experimental probabilities.

   (1) Toss a paper cup and observe how it lands: open-end down, open-end up, or on its side.

   (2) Spin a coin and observe how it stops: heads up, tails up, or on its side.

   (3) Roll a number cube that is slightly weighted on one side.

   (4) Choose one card randomly from a deck of cards where some cards are sticking to each other. Observe whether you get a diamond. Put the card back into the deck in a random location before choosing the next card.

   (5) Choose one card randomly from a deck of cards and observe whether you get a diamond. If it is *not* a diamond, put the card back randomly into the deck. If it *is* a diamond, place the card at the *back* of the deck. Don't reshuffle the deck at any point.

   (6) Put several stuffed animals in a hat or box. The animals should vary in size as much as possible. Each time, pull out one animal randomly, then put it back.

   If you would like, or if your teacher so decides, you could do more than one of these experiments.

Outcome	Relative Frequency	Experimental probability (%)
	____/100	

Counting the Possibilities

A **sample space** is a list of all possible outcomes of an experiment.

Example 1. We roll two dice. The sample space for this experiment is shown in the grid on the right. Each dot represents one outcome. For example, the point (1, 4) means that the first die shows 1 and the second die shows 4.

Notice that there are a total of 6 · 6 = 36 possible outcomes.

What is the probability of getting the sum of 8 when rolling two dice? The chart helps answer that question. First we find out and count how many outcomes give you the sum 8:

You could roll 2 + 6, 3 + 5, 4 + 4, 5 + 3, or 6 + 2. Those number pairs are circled in the second graphic.

So there are five favorable outcomes out of 36 possible. Therefore, the probability of getting 8 as a sum is 5/36.

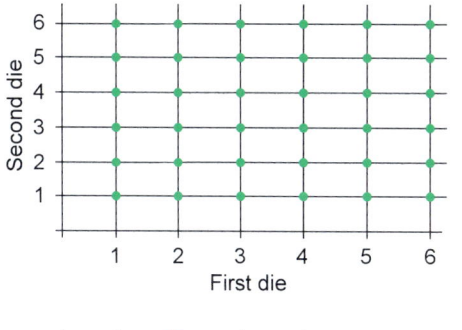

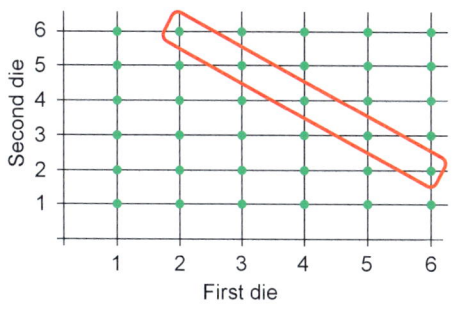

1. **a.** How many outcomes are there for rolling the same number on both dice (such as (5, 5))?

   b. What is the probability of rolling the same number on both dice?

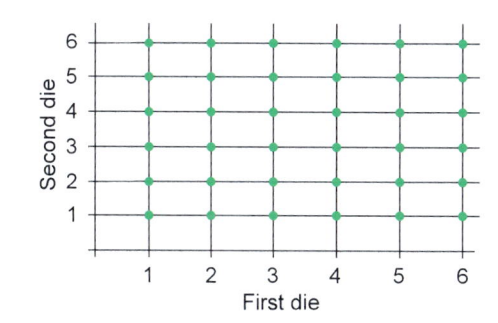

2. **a.** What is the probability of rolling 5 on the first die and 6 on the second?

   b. What is the probability of rolling 5 on one die and 6 on the other?

   c. What is the probability of getting a sum of 7 when rolling two dice?

   d. What is the probability of getting a sum of at least 6 when rolling two dice?

3. You roll a six-sided die twice. Find the probabilities.

   a. P(1; 5)

   b. P(2; 5 or 6)

   c. P(even; odd)

   d. P(6; not 6)

21

The array we used on the previous page can show the sample space (all the possible outcomes) for only two events, like rolling two different dice. A **tree diagram** can show more than two events, so it is a common way to represent the sample space for multiple events.

Example 2. Peter has white, blue, yellow, and red shirts, blue and white slacks, and brown and blue tennis shoes. How many possible ways can he make an outfit using them?

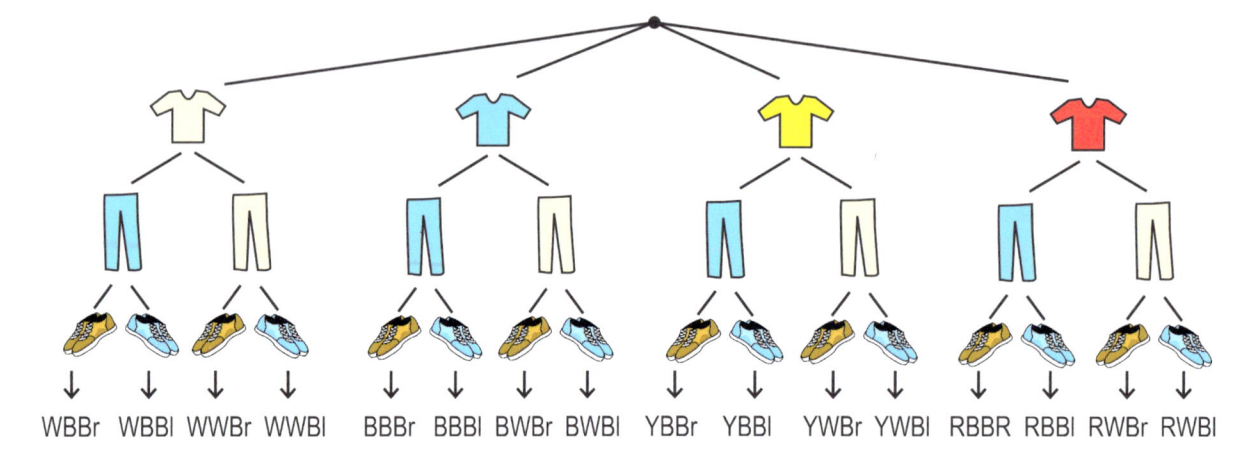

WBBr WBBl WWBr WWBl   BBBr BBBl BWBr BWBl   YBBr YBBl YWBr YWBl   RBBR RBBl RWBr RWBl

At the bottom we have listed all the possible outcomes using letter combinations. This is optional, but helpful. For example, WBBr means a white shirt, blue slacks, and brown shoes.

Notice that in the first level, there are 4 possibilities, in the second level there are 2 possibilities, and in the last level 2 possibilities. In total, there are 4 × 2 × 2 = 16 ways he can make an outfit.

Example 3. Peter chooses his shirt, slacks, and shoes randomly. What is the probability that his shirt and slacks match?

"Matching" means that he wears a white shirt with white pants or a blue shirt with blue pants. Since the shoes are not specified, there are four possible outfits: WWBr, WWBl, BBBr, or BBBl. So the probability is 4/16 = 1/4.

Example 4. What is the probability that Peter wears a red shirt?

You can count the four outfits with a red shirt to get the probability as 4/16 = 1/4. But it's simpler to ignore the other clothing items and just look at the shirts: the red shirt is one of the four, so the probability is 1/4.

4. **a.** Complete the tree diagram to show the outcomes when you first roll a die, then toss a coin. The bottom row lists the outcomes using number-letter combinations, such as 1H and 1T.

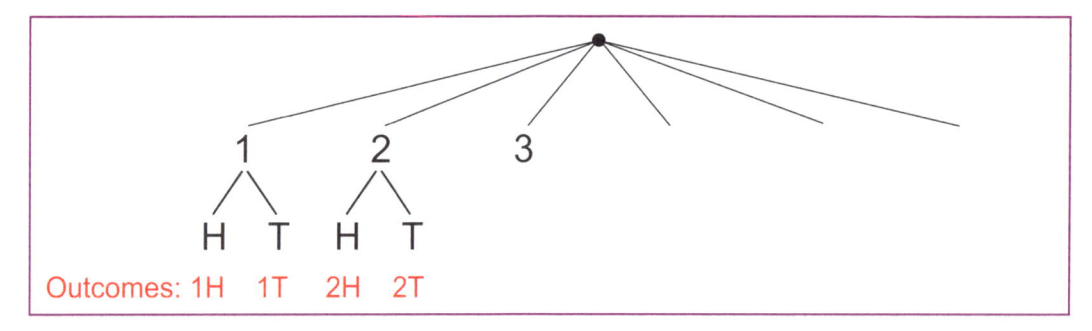

Outcomes: 1H 1T 2H 2T

Now find these probabilities:

b. P(even number, heads)

c. P(not 6, heads)

d. P(4 or more, tails)

e. P(any number, tails)

5. A restaurant offers the following menu:

   Second course: soup or salad

   Main course: fish, chicken, or beef

   Dessert: ice cream or cake

   a. Cindy chooses her second course, main course, and dessert randomly. Draw a tree diagram for the sample space.

   What is the probability that Cindy ...

   b. ... gets ice cream for dessert?

   c. ... gets soup, fish, and cake?

   d. ... eats soup and fish (and either dessert)?

   e. ... eats salad and ice cream?

   f. ... doesn't eat chicken?

   g. ... doesn't eat fish or ice cream?

6. You take a marble out of the bag and *put it back*. Then you take another marble out. Complete the table that lists the sample space (all the possible outcomes). Notice that we have to list both red marbles and both green marbles separately.

Second marble → First marble ↓	R	R	G	G	B
R	RR	RR	RG	RG	RB
R	RR				
G	GR				
G	GR				
B	BR				

Now find the probabilities:

a. P(red, then green)

b. P(green, then red)

c. P(not blue, not blue)

d. P(not red, not red)

23

7. You take a marble out of this bag *without* putting it back, and then you take another marble. In effect, you take two marbles out of the bag.

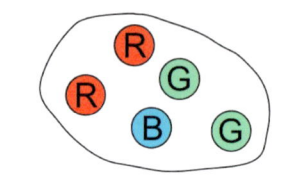

   a. Complete the tree diagram for this experiment. Notice: which marble you take out determines which marbles are left. For example, if the first marble is red, then the bag has 1 blue, 2 green, and *only* 1 red marble left to choose from.

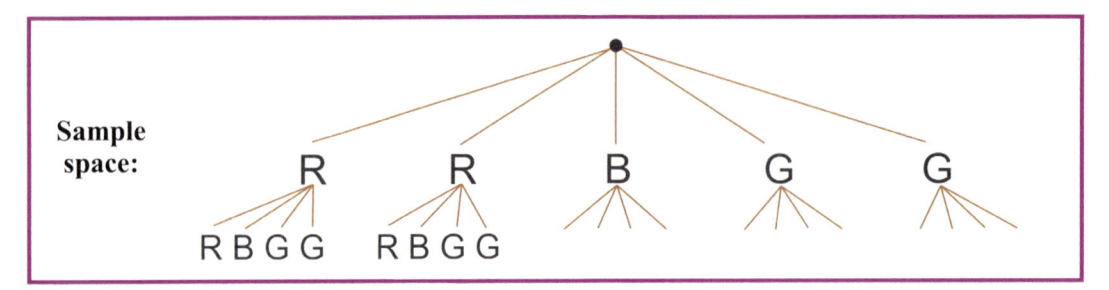

   Sample space:

   R → R B G G
   R → R B G G
   B →
   G →
   G →

   Then find the probabilities:

   b. P(not red, not red)

   c. P(red, then green)

   d. P(green, then red)

   e. Add the probabilities from (c) and (d) to get the probability of choosing exactly one red and one green marble, in either order.

   f. (Optional) Conduct this experiment. If you don't have marbles, you could let red = quarters, green = dimes, and blue = nickels, and perform the experiment with coins. Observe for example whether the probability you calculated in (e) for getting one red and one green marble is close to what you observe in your experiment.

8. You make a two-digit number by choosing both digits randomly from the numbers on the cards. The card is replaced after each choice.

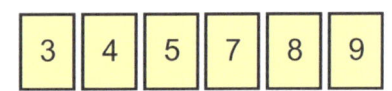

   | 3 | 4 | 5 | 7 | 8 | 9 |

   Sample space:

   3,3   3,4   3,5   3,7   3,8   3,9

   a. In the space on the right, finish listing all the possible outcomes of this experiment.

   Use the list to find these probabilities:

   b. P(4; 9)

   c. P(even; 7)

   d. P(even; odd)

   e. P(less than 6; more than 6)

   f. P(not 6; not 6)

   g. P(both digits are the same)

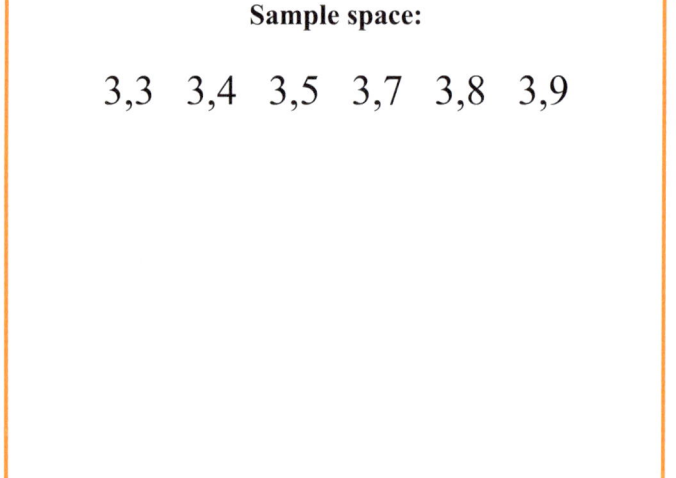

9. A special education classroom has 4 boys and 2 girls. The teacher randomly chooses two students to be responsible for the cleanup after a bake sale.

   a. Make a tree diagram for the sample space. Notice that if the first student is a girl, then there are 4 boys and 1 girl left to choose the second student from. If the first student is a boy, then there are 3 boys and 2 girls left to choose the second student from.

   Now use the sample space and give these probabilities as fractions.

   b. What is the probability that both students are girls?

   c. What is the probability that both students are boys?

   d. What is the probability that the first student chosen is a girl, and the second is a boy?

   e. What is the probability that the first student chosen is a boy, and the second is a girl?

   Now check. The probabilities you get in (b), (c), (d), and (e) should total 1 because they are all the possible outcomes.

   f. Add the probabilities in (d) and (e) to get the probability that one of the cleaners is a girl and one is a boy.

10. In tossing two distinct coins, one of the possible outcomes is HT: first coin heads, second coin tails.

   a. List all the possible outcomes.

   b. Each of the possible outcomes is equally likely. Therefore, what is the theoretical probability of each outcome?

   c. Now toss two coins 200 times and compare the experimental probabilities to the theoretical ones. Before you do, predict about how many times you would expect to see each outcome:

   _____ times

   Note: You need to distinguish the coins somehow: Either use different coins, like a 5 c and a 10 c, or mark identical coins in some manner, maybe as "1" and "2." Distinguishing the coins is necessary because the outcomes HT and TH aren't the same. You need to know which is which.

   You may also run the simulation in the spreadsheet file #2 listed at
   https://www.mathmammoth.com/lessons/probability_simulations.php

   If you have the digital version of the curriculum, the spreadsheet file is included in your download.

Outcome	Frequency	% of total tosses
TOTALS	50	100%

   d. Check whether the observed frequencies are fairly close to those predicted by the theoretical probabilities.

   Let's say they were not. What could be the reason?

Puzzle Corner

In a multiple-choice test, you have four choices (a, b, c, and d) for your answer each time.

a. Let's say the test has two questions and Andy chooses both answers randomly. What is the probability that Andy gets both questions correct?

b. Let's say the test has *five* questions and Kimberly answers them all randomly. What is the probability she gets them all correct?

Using Simulations to Find Probabilities

1. Let's say that a reporter interviews ten people at random on the street. If the probability that he selects a male or a female is 50/50, then what is the probability that, of the ten people, exactly 5 are male and 5 are female?

   We will use a *simulation* to answer this question.

   If you have the download version of this curriculum, you can use the included spreadsheet file, which simulates choosing 10 people by generating sets of 10 random numbers (0s and 1s).

   Another possibility is to use the virtual coin tossing tool at https://syzygy.virtualave.net/multicointoss.htm Choose 10 coins and 1 set of coin tosses. Record the outcome (how many males/females). Repeat your experiment at least 100 times, recording each outcome.

   Yet a different way to do the simulation is to toss 10 coins 100 times yourself or with the help of others.

   https://syzygy.virtualave.net/multicointoss.htm

   A result of four heads and six tails simulates interviews with four females and six males.

   a. Fill in the table on the right based on the simulation.

   b. Based on your simulation, what is the experimental probability that exactly 5 of the randomly chosen people are female and 5 of them are male?

Outcome	Frequency	Experimental probability
0 F 10 M		
1 F 9 M		
2 F 8 M		
TOTALS		100%

   c. What is the probability that 4 of the people are female and 6 are male?

   d. What is the probability of getting at least 3 females *and* at least 3 males in the set of 10 people?

   Hint: There are several possible outcomes with at least 3 females and at least 3 males.

   e. What is the probability of getting only 1 or 2 of either sex (and 9 or 8 of the other)?

27

2. The chance that a random student at Lazyville High School has completed homework on time is 50%. One day, the principal chooses six of the school's students at random. Design a simulation to study this situation.

Explain your design:

Now run the simulation. Repeat the experiment at least 100 times, but more is better. Record each outcome. Then count how many of your outcomes represent zero students finishing homework on time, one student finishing it on time, and so on, and calculate the corresponding probabilities.

Results of simulation		
Students who finished homework	Relative Frequency	Experimental probability
0		
1		
2		
3		
4		
5		
6		
TOTALS		100%

Now use the results of the simulation to answer the following questions:

a. What is the probability that two of the six students have completed homework on time?

b. What is the probability that only one student has completed homework on time?

c. What is the probability that none of the six completed homework on time?

d. What is the probability that *at most* 2 of them have completed homework on time?
   Hint: Add the probabilities from (a), (b), and (c).

e. What is the probability that *at least* 3 of them have completed homework on time?

3. Let's say that the students at Lazyville High School have changed for the better and that the chance that a random student has completed homework on time is now 70%. Once again, the principal chooses six students at random. What is the probability that at least three of them have completed homework on time?

This time, we cannot use coin tosses to simulate this experiment because the probability that a single student has completed homework on time is no longer 50%, but we can still use simulation. Make up a deck of cards so that 70% of them are diamonds. For example, make a deck with 7 diamonds and 3 other cards. Then choosing one card randomly represents choosing a student randomly.

Now choose one card randomly from your deck and put it back. Repeat this six times. The six choices represent *one* outcome of the principal choosing six students). Record the outcome, for example as DXDDXD (D = diamond; X = other card).

Now repeat this experiment (choosing one card 6 times) at least 50 times (more is better). Record each outcome. Then count how many of your outcomes represent zero students finishing homework on time, one student finishing it on time, and so on, and calculate the corresponding experimental probabilities.

Students who finished homework	Relative Frequency	Experimental probability
0		
1		
2		
3		
4		
5		
6		
TOTALS		100%

Use the results of the simulation to answer the following questions:

a. What is the probability that at most 2 of them have completed homework on time?

b. What is the probability that at least 3 of them have completed homework on time?

c. What is the probability that not all of them have completed homework on time?

d. What is the probability that at least two-thirds of them have completed homework on time?

Example 1. Choosing one card from a deck hundreds of times, like you did in exercise 3, is a lot of work. An easier approach is to use computer-generated random digits, such as from **https://random.org/integers**

For the situation in exercise 3, you can generate integers between 0 and 9. To get the 70% probability, let the integers from 0 to 6 represent completing homework on time and 7 to 9 represent not completing homework on time.

Part 1: The Integers

Generate 600 random integers (maximum 10,000).

Each integer should have a value between 0 and 9

Format in 6 column(s).

Generating 600 integers and formatting the results in 6 columns will get you 100 rows with six numbers in each. Each row of six numbers represents one repetition of choosing of 6 students randomly.

For example, in the row 6 1 1 1 9 9, four out of the six numbers are 6 or less.
This represents choosing a set of six students where four of them had completed homework on time.

If you don't have Internet access, you can create random numbers in a spreadsheet program. For example, in Excel, type the formula =FLOOR(RAND()*10, 1) into a cell.

RAND() creates a random decimal number between 0 and 1. Then we multiply it by 10 and round it down to the nearest whole number (the "floor") in order to get a random number between 0 and 9.

Lastly, to get a sequence of random numbers, copy that formula to a large number of cells.

4. Repeat exercise 3 using random digits. Generate at least 200 repetitions of choosing six students. Here's the situation again: The chance that a random student from Lazyville High School has completed homework on time is 70%. The principal chooses six students at random.

Students who finished homework	Relative Frequency	Experimental probability
0		
1		
2		
3		
4		
5		
6		
TOTALS		100%

a. What is the probability that at least half of them have completed homework on time?

b. What is the probability that at least four out of the six have completed homework on time?

5. A certain health care center typically gets 20 people per day who come to donate blood. Forty percent of them have type A blood. We are interested in determining how many donors we'll need to check before we find one with type A blood

   This question can be framed more exactly in terms of probabilities. For example, we could ask:

   - What is the probability that the first donor of the day has type A blood?
   - What is the probability that the first donor of the day *doesn't* have type A blood but the second one does? Let's denote this event as XA, where X means a person with some other blood type and A means a person with blood type A.
   - What is the probability of the event XXA — that only the third donor of the day has type A blood?
   - What is P(XXXA) — that it will take 4 donors to find one with type A blood?
   - What is P(XXXXA) — that it will take 5 donors to find one with type A blood?

   And so on.

   a. Design a simulation to study these types of questions. Explain your design here:

   b. Let's look at just the first four donors. Limit your simulation to a sequence of four donors, and repeat it at least 100 times. More is better. Record each outcome. Then count the frequencies of each listed event to complete the table.

   Note: in the table, the "_" denotes a person with any blood type, including type A. For example, the event "AXAA" is classified as part of the event "A _ _ _".

Results of simulation		
Event	Frequency	Experimental probability
A _ _ _		
XA _ _		
XXA _		
XXXA		
XXXX		
TOTALS		100%

   Find the probabilities:

   c. What is the probability that it will take 1, 2, or 3 donors until you find one with blood type A?

   d. What is the probability that it will take exactly 4 donors until you find one with blood type A?

   e. What is the probability that it will take more than 4 donors until you find one with blood type A?

   f. What is the probability that it will take at least 4 donors until you find one with blood type A?

6. An ideal gumball machine with an unending supply of gumballs dispenses them in four different flavors: strawberry, lemon, blackberry, and apple. Each flavor is equally likely.

   a. You get *three* gumballs out of the dispenser (randomly). Design a simulation for this experiment. Explain your design here:

   b. Run your simulation for at least 100 repetitions (but more is better). Record the outcomes.

   Hint: If you generate the random numbers at https://random.org/integers, copy the rows of random numbers to a spreadsheet program, such as Excel. You may need to use the "Paste special" command to paste them as "text." Then you can sort them easily. Simply choose all the cells you want to sort and then find "Sort" in the program's commands. In Excel it is "Data → Sort." Then choose the column by which to sort them. For example, if you want to study the first gumball in the set of three, sort the data by the first column.

   Find the experimental probabilities:

   c. P(none are strawberry) =

   d. P(exactly 2 are strawberry) =

   e. P(all 3 are strawberry) =

   f. P(none are lemon or blackberry) =

   g. P(all three are the same flavor) =

Probabilities of Compound Events

(This lesson is optional.)

Up to now we've been looking only at **simple events**, events that require just a single calculation of probability. A **compound event** is an event that consists of two or more simple events. If the outcome of one event does not affect the outcome of another, the events are said to be **independent**. If the compound event consists only of independent simple events, then it is very easy to calculate the probability of the compound event: we simply multiply the probabilities of the individual simple events. The examples will make this clear.

Example 1. You roll a die and toss a coin. What is the probability of rolling a 6 and getting heads?

P(6) is 1/6 and P(heads) is 1/2. Clearly, whether you get heads or tails on the coin does not affect what you get on the roll, so the two events are independent. Therefore, we can multiply the two probabilities:

$$P(6 \text{ and heads}) = \frac{1}{6} \cdot \frac{1}{2} = \frac{1}{12}$$

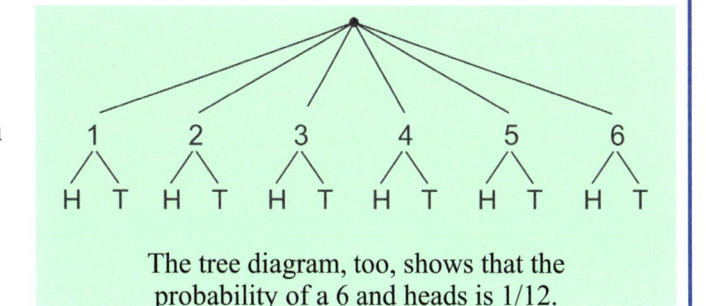

The tree diagram, too, shows that the probability of a 6 and heads is 1/12.

Example 2. You toss a coin three times. What is the probability of getting heads every time?

These three events — toss a coin, toss a coin, toss a coin — are independent. Getting heads on one toss doesn't affect whether you get heads or tails on the next.

P(heads) = 1/2. Therefore, P(heads and heads and heads) = $\frac{1}{2} \cdot \frac{1}{2} \cdot \frac{1}{2} = \frac{1}{8}$.

You can also see this result in the tree diagram. Only one outcome out of the 8 is "HHH."

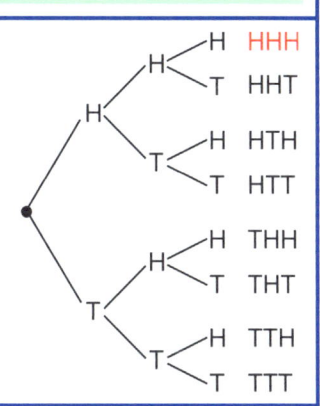

Example 3. A bag has three red marbles, two blue marbles, and five green marbles. You take out a marble and put it back. Then you take out a marble again and put it back. What is the probability of getting first a red marble and then a blue one?

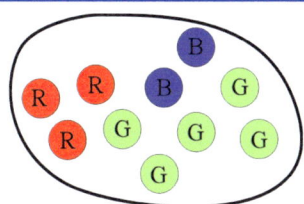

Again, we simply multiply the individual probabilities:

$$P(\text{red, blue}) = \frac{3}{10} \cdot \frac{2}{10} = \frac{6}{100} = \frac{3}{50}.$$

1. You toss a coin three times.

   a. What is the probability of getting tails, then heads, then tails?

   b. What is the probability that you get heads on your second toss?

   c. Use the tree diagram in Example 2. What is the probability of getting heads twice and tails once in three tosses? Note that they can be in any order, such as THH or HTH.

2. You take a marble out of the bag and put it back. Then you take out another marble. Find the probabilities:

   a. P(red, then green)

   b. P(green, then red)

   c. P(not blue, not blue)

   d. P(not green, not green)

3. Helen is a teacher. She has eight different outfits that she wears to school, and each day she chooses her outfit randomly from among those eight. One of her outfits is pink, and her students don't like it.

   a. What is the probability that she does *not* wear the pink outfit for five days in a row?

   b. What is the probability that she does not wear the pink outfit for ten days in a row?

4. You make a two-digit number by choosing the digits randomly using these number cards. (You put the card back after choosing.)

   a. How many different numbers is it possible to make?

   b. What is the probability of making the number 35?

   c. What is the probability of making a number that is divisible by 9?

5. The weatherman says that the chance of rain is 20% for each of the next five days, and your birthday is in two days! You also know that the probability of your dad taking you to the amusement park on your birthday is 1/2.

   a. What is the probability that you get to go to the park, and it doesn't rain?

   b. What is the probability that you get to go to the park, and it rains?

   Check: The sum of the probabilities in (a) and (b) should be 1/2.

Taking an object without replacing it

Example 4. You choose a marble from this bag and don't put it back. Then you choose another marble. This means you have in effect chosen two marbles. What is the probability that both are red?

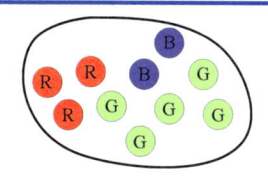

We first find the probability that the *first* marble is red. That is simply 3/10 since there are three red marbles and ten in all.

After you get a red marble and don't put it back, the bag now has only two red marbles and nine in total. So the probability of getting a red marble now is 2/9.

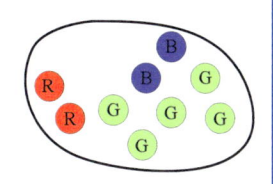

We multiply the two probabilities to find the probability of getting two red marbles:

$$P(\text{red and red}) = \frac{\cancel{3}^1}{\cancel{10}_5} \cdot \frac{\cancel{2}^1}{\cancel{9}_3} = \frac{1}{15}$$

6. You choose one card without putting it back. Then you choose another. What is the probability that the first is an even number and the second is 7? Fill in Joanne's solution to this question.

    At first, there are four cards with an even number and eight cards in total to choose from.

    So P(even) = _____.

    After one card with an even number has been drawn, there are seven cards left, and one of them is 7.

    So P(7) is _____.

    Then we multiply the two probabilities to get the probability of both events:

    P(even, 7) =

7. You choose a card randomly from this set of cards. Then you choose another card (without replacing the first). Find the probabilities.

    a. P(heart, heart)

    b. P(star, cross)

    c. P(not heart, not heart)

    d. P(star, not star)

8. You choose two marbles randomly from the bag without replacing them.

    a. What is the probability that both are green?

    b. What is the probability that neither is green?

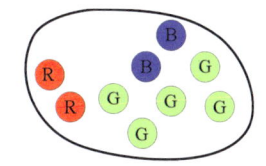

    c. What is the probability that exactly one of the marbles is green?
    Hint: Do you want either of the events in (a) or (b) to occur?

35

9. Michael has 10 white socks and 14 black socks mixed together in a drawer. He randomly chooses one sock to wear and doesn't put it back. Then he chooses another sock. Find the probabilities:

   a. P(white, white)

   b. P(black, black)

   c. P(black, white)

   d. P(white, black)

   CHECK. The four probabilities above should total 1 or 100%.

   e. Add the probabilities in (a) and (b) to find the probability that Michael wears matching socks.

   f. What is the probability Michael *doesn't* wear matching socks?

10. You choose three cards from a standard 52-card deck (don't include the jokers). Calculate the following probabilities to the nearest tenth of a percent:

    a. What is the probability that all of them are aces (AAA)?

    b. What is the probability that the first of the three cards is an ace and the others are not (AXX)?

    c. What is the probability that the first of the three cards is not an ace, the second is, and the third is not (XAX)?

    d. What is the probability for XXA?

    e. Add the three probabilities from (b) through (d) to get the probability that exactly one of the three cards is an ace.

Puzzle Corner

Matthew has 8 white socks, 9 brown socks, and 10 black socks mixed together in a drawer. He chooses two socks randomly. Find the probability, to the nearest hundredth of a percent, that he gets to wear a matching pair.

Review

1. The chart lists the favorite school subjects of the students in a 7th grade classroom.

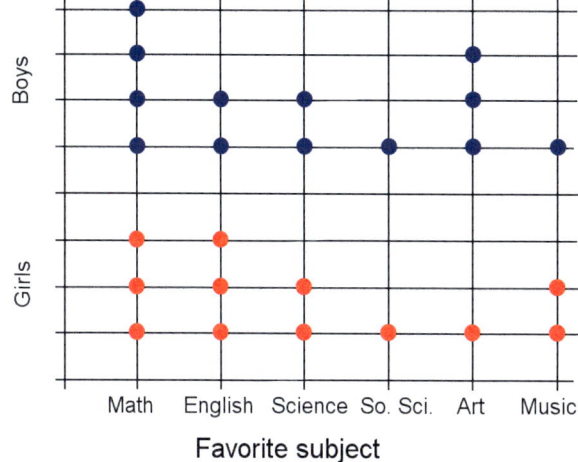

    a. You choose one student randomly. What is the probability that the student's favorite subject is not math, English, or science?

    b. What is the probability that a randomly chosen student's favorite subject is math?

    c. Now look at the boys only. If you choose one boy randomly, what is the probability that his favorite subject is math?

    d. If you choose a girl randomly from among the girls, then what is the probability that her favorite subject is math?

2. Abigail rolls two dice. Find the probabilities of these events:

    a. P(5, 6)

    b. P(even, even)

    c. P(at least 5, at least 5)

    d. P(at most 3, at most 2)

3. Julie and Jane experimented with rolling a die. They rolled the die 60 times in a row and recorded the results:

    2 4 3 3 5   2 5 4 1 6   1 4 4 4 5   4 2 6 4 1   5 2 1 1 1   2 1 3 6 2
    4 2 2 3 3   5 2 3 6 4   4 4 1 3 2   6 6 4 5 6   4 6 5 6 5   6 5 6 5 3

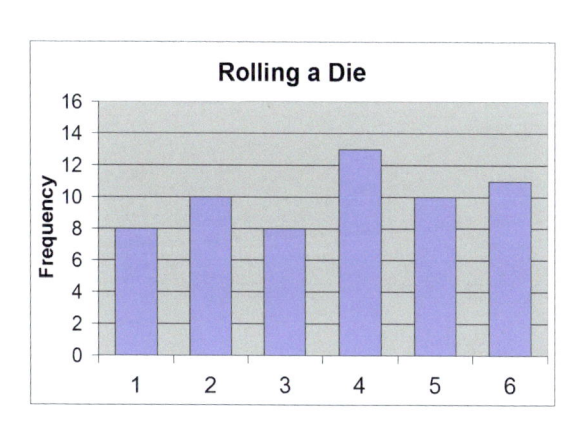

    a. In this experiment, what was the probability of rolling a 1?

    Rolling a 4?

    b. Why are those probabilities not 1/6?

4. Andrew did an experiment where he tossed two coins 200 times and recorded the outcomes. The table below shows his results. "H" means "heads," and "T" means "tails," so "HH," for example, means both coins landed "heads."

   a. Calculate the experimental and theoretical probabilities and fill in the table.

Outcome	Frequency	Experimental Probability (%)	Theoretical Probability (%)
HH	38		
HT	53		
TH	46		
TT	63		
TOTALS	200		

   b. How would the experimental probabilities change if Andrew redid this experiment with 2,000 tosses?

5. In a multi-player computer game, a computer chooses colors randomly for a girl's dress. The first color it chooses is the main color of the dress. The second color is for the bows and some layers of the skirt. The computer uses this list of colors: *red, blue, purple, pink, orange, yellow, mint.*

   After choosing the main color, the computer removes it from the list and chooses the second color from the resulting list of six colors. That way the dress is sure to have two different colors.

   a. What is the probability the computer chooses first purple, then orange?

   b. What is the probability the computer chooses first red, then not pink?

   c. Janet doesn't like mint. What is the probability her character gets a dress with no mint in it when she plays the game?

Probability, Grade 7 Answer Key

Probability, p. 13

1. The possible outcomes are: red marble, blue marble, green marble.

2. Since there are ten marbles, the probability of choosing any individual marble is 1/10.
    a. There are two blue marbles, so P(blue) = 2/10 = 1/5.
    b. There are three red marbles, so there are seven that aren't. Thus P(not red) = 7/10.
    c. If the marble is neither blue nor green, then it must be one of the three red ones: P(not blue or green) = 3/10.

3. Answers will vary. Check the student's answer. For example: pick a purple marble.

4. a. The possible outcomes are P, R, O, B, A, I, L, T, and Y.
    b. P(B) = 2/11.
    c. P(A or I) = 3/11.
    d. P(vowel) = 5/11. (In PROBABILITY, the Y is a vowel.)
    e. Answers will vary. Check the student's answer. The probability of the student's event should be quite a bit more than ½ and less than 1. For example, choosing "the probability that the letter comes after H in the alphabet" would work because all but three of the letters come after H in the alphabet, so P(letter after H) = 8/11.

5. Since P(raining) = 1/10, P(not raining) 1 − 1/10 = 9/10.

6. a. P(green) = 2/8 = 1/4.
    b. P(not green) = 6/8 = 3/4.
    c. P(not pink) = 7/8.
    d. P(not black) = 8/8 = 1.
    e. Answers will vary. Check the student's answer. The probability of the student's event should be more than 0 and quite a bit less than ½. For example: "the spinner lands on brown" fits, because P(brown) = ⅛.

7.

Possible outcomes	Probability (fraction)	Probability (percentage)
heart	1/3	33.3%
star	2/9	22.2%
cross	4/9	44.4%

8. a. P(yellow) = 1/6 = 16.7%.
    b. P(blue or green) = 2/6 = 33.3%.
    c. P(not orange) = 5/6 = 83.3%.
    d. P(not red and not purple) = 4/6 = 66.6%.
    e. Answers will vary. Check the student's answer. The probability of the student's event should be 1. For example, "the spinner lands on a rainbow color" is a sure event, and its probability is 1.

9. a. P(window seat) = 22/45 = 48.9%
    b. P(window seat) = 17/37 = 45.9%

Puzzle Corner. Of the 45 seats on the bus, 15 are occupied. So 45 − 15 = 30 are unoccupied. There are 22 window seats, so at least some of them must be unoccupied. The probability of getting a window seat is:

P(window seat) = (unoccupied window seats/all unoccupied seats) = (unoccupied window seats)/30.

We don't know the number of unoccupied window seats, yet the driver said that the probability of getting one at random is less than 25%. So (unoccupied window seats)/30 < 25%. To solve that you can simply guess and check. For example, if you guess that 6 people are sitting in window seats, then 9 people aren't. In that case, the probability of getting a window seat is: (unoccupied window seats)/30 = 16/30 ≈ 53.3%, which is too big.

Guess-and-check solution: We know the probability of getting a window seat is less than 25%. If there were 8 unoccupied window seats, the probability of getting a window seat would be 8/30 ≈ 26.7%. If seven were unoccupied, the probability would be 7/30 ≈ 23.3%. So, at most, seven window seats are unoccupied, which means that at least 15 window seats are occupied. Of course this means that all 15 people in the bus are seated in window seats!

Probability, cont.

Algebra solution: Of the 45 total seats, only 30 are available. Let w be the number of unoccupied window seats. Then the probability of getting a window seat is w/30, which is less than 25% or 0.25, so we get the inequality w/30 < 0.25. If we multiply both sides by 30, we get w < 7.5. Since we're not talking about parts of a seat, the solution we want is w = 7. So out of 30 unoccupied seats, 7 are window seats, which gives us a probability of choosing one at random that is less than 25%, as required. And 22 − 7 =15 window seats are already occupied.

Probability Problems from Statistics, p. 16

1. a. It is easier to calculate the probability of the complement event, which is that the student's score was at most D−.
   P(at most D−) = (3 + 3)/70 = 6/70. Then, P(at least D) = 1 − 6/70 = 64/70 = <u>32/35</u>.

   b. It is easier to calculate the probability of the complement event, which is that the student's score was at least A−.
   P(at least A−) = (2 + 4)/70 = 6/70. Then, P(at most B+) = 1 − 6/70 = 64/70 = <u>32/35</u>.

2. a. It is easier to calculate the probability of the complement event, which is that the child is at least 10 years of age.
   P(at least 10) = 2/20. Then, P(at most 9) = 1 − 2/20 = <u>18/20 = 9/10</u>.

   b. It is easier to calculate the probability of the complement event, which is that the child is at most 6 years of age.
   P(at most 6) = 3/20. Then, P(at least 7) = 1 − 3/20 = <u>17/20</u>.

3. P(English) = 215,423,557/262,375,152 = 82.11%. Then, P(not English) = 100% − 82.11% = <u>17.89%</u>.

4. a. P(female 25-29 yrs old) = 9,582,576/281,421,906 = <u>3.4%</u>.

   b. P(female 25-29 yrs old) = 9,582,576/143,368,343 = <u>6.7%</u>.

   c. First let's calculate the number of males who are at most 14 years old:
   9,810,733 + 10,523,277 + 10,520,197 = 30,854,207 males. There are a total of 138,053,563 males, so the number of males who are at least 15 is 138,053,563 − 30,854,207 = 107,199,356.

   Now, P(random person is a male at least 15 years old) = 107,199,356/281,421,906 = <u>38.1%</u>.

   d. The number of persons who are at least 65 is 9,533,545 + 8,857,441 + 7,415,813 + 4,945,367 + 4,239,587
   = 34,991,753. The number of persons who are at most 64 is then 281,421,906 − 34,991,753 = 246,430,153.

   Then, P(at most 64 years) = 246,430,153/281,421,906 = <u>87.6%</u>.

Experimental Probability, p. 18

1. a. No, the various outcomes (1 through 6) will not occur 10 times. Results will vary. Check the student's results. See the table below for an example.

   b. How many times would we expect to roll each number, based on the theoretical probabilities of 1/6? <u>20</u> times
   Results will vary. Check the student's results. See the table below for an example.

   c. Results will vary. Check the student's results. See the table below for an example.

a. With 60 rolls:		
Outcome	Frequency	Probability
1	7	11.67%
2	13	21.67%
3	8	13.33%
4	9	15.00%
5	11	18.33%
6	12	20.00%
TOTALS	60	100%

b. With 120 rolls:		
Outcome	Frequency	Probability
1	21	17.50%
2	19	15.83%
3	19	15.83%
4	25	20.83%
5	14	11.67%
6	22	18.33%
TOTALS	120	100%

c. With 480 rolls:		
Outcome	Frequency	Probability
1	92	19.17%
2	92	19.17%
3	67	13.96%
4	78	16.25%
5	67	13.96%
6	84	17.50%
TOTALS	480	100%

Experimental Probability, Cont.

1. d. The experimental probabilities will be closest to the theoretical ones (16.67%) for 480 rolls. However, it's possible, though not likely, that the student's results will not show that. It can take many hundreds of rolls before the experimental probabilities start getting within even a few percentage points of the theoretical 16.67%. For 60 and even 120 rolls, experimental results can vary wildly from the theoretical 16.67%.

2. a. Predict about how many times you expect to get each of the six possible numbers if you roll a die 1,000 times:

   About 167 times

   b. Results will vary. Check the student's results. See the table on the right for an example.

   Each experimental probability should be fairly close to the theoretical probability of 1/6 = 16.67%, typically within about 2 percentage points, though it could vary more than that.

   c. The experimental probabilities should be closer to 16.67% than what they were when rolling a die 1,000 times.

   d. Results will vary; probably somewhere between about 15,000 and 20,000 rolls.

Outcome	Frequency	Experimental Probability (%)
1	171	17.1%
2	156	15.6%
3	159	15.9%
4	161	16.1%
5	172	17.2%
6	181	18.1%
TOTALS	1000	100%

3. Results will vary a lot since the composition of the 12 cards will vary and since the exercise involves a chance process. Check the student's results. See the table below for an example using the cards 2, 2, 2, 3, 3, 4, 5, 5, 5, 5, 6, 6.

   a. See column 3 of the table.

   b. See columns 2 and 4 of the table. The theoretical and experimental probabilities should be relatively close. However, they might not be if the process for drawing cards was not truly random. Each card has to have an equal chance of being drawn, and that includes the first and the last cards of the set. For example, if the person drawing the card puts it back at the back, yet always draws a new card from the middle, that is not a truly random process of drawing.

Outcome	Theoretical probability	Relative Frequency	Experimental probability
2	25%	28/100	28%
3	16.67%	17/100	17%
4	8.33%	7/100	7%
5	33.33%	31/100	31%
6	16.67%	17/100	17%

Another example of a non-random drawing is if the card is always put back in the middle and the person drawing cards tends to draw cards from the middle more often than near the ends.

In fact, it may be difficult to come up with a way to draw a card from a deck truly at random, because people tend to want to draw cards from the middle. It might be better to put the 12 cards into a hat and then let someone draw one.

4. Results will vary. Check the student's results.

Counting the Possibilities, p. 21

1. a. There are six outcomes: (1, 1), (2, 2), (3, 3), (4, 4), (5, 5), and (6, 6).
   b. P(doubles) = 6/36 = 1/6

2. a. There is only one such outcome: P(5, 6) = 1/36.

   b. There are two favorable outcomes: (5, 6) and (6, 5) so the probability is 2/36 = 1/18.

   c. The favorable outcomes are (1, 6), (2, 5), (3, 4), (4, 3), (5, 2), and (6, 1). The probability is 6/36 = 1/6.

   d. It is easier to find the outcomes for the sum that is 5 or less: (1, 1), (1, 2), (1, 3), (1, 4), (2, 1), (2, 2), (2, 3), (3, 1), (3, 2), (4, 1), which is 10 outcomes. This means that in 26 outcomes the sum is 6 or more. The probability is therefore 26/36 = 13/18.

Counting the Possibilities, Cont.

3. a. This situation is identical to rolling two dice, so we can use the same chart for the sample space. P(1; 5) = 1/36.

   b. The favorable outcomes are (2, 5) and (2, 6). The probability is therefore 2/36 = 1/18.

   c. The favorable outcomes for getting an even number on the first die and an odd number on the second are: (2, 1), (2, 3), (2, 5), (4, 1), (4, 3), (4, 5), (6, 1), (6, 3), and (6, 5), a total of nine possible outcomes. Therefore, P(even; odd) = 9/36 = 1/4.

   d. The favorable outcomes are (6, 1), (6, 2), (6, 3), (6, 4), and (6, 5), which is a total of five outcomes. P(6; not 6) = 5/36.

4. a.

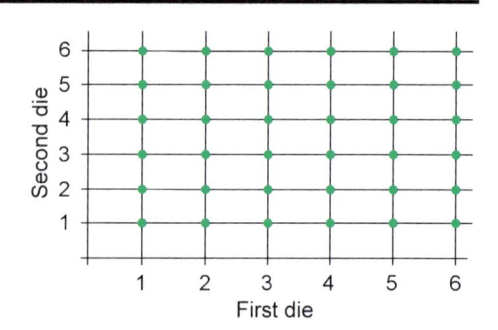

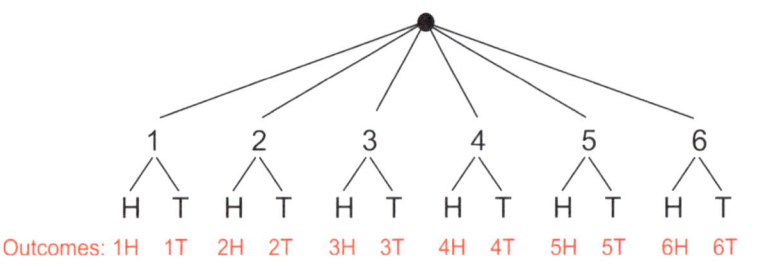

   ```
                    •
         / / /  |  \ \ \
        1  2  3  4  5  6
       /\ /\ /\ /\ /\ /\
       H T H T H T H T H T H T
   Outcomes: 1H 1T 2H 2T 3H 3T 4H 4T 5H 5T 6H 6T
   ```

   b. There are three favorable outcomes — 2H, 4H, and 6H — out of a total of 12 outcomes. So P(even number, heads) = 3/12 = 1/4.
   c. The favorable outcomes are: 1H, 2H, 3H, 4H, 5H. So P(not 6, heads) = 5/12.
   d. The favorable outcomes are: 4T, 5T, 6T. So P(4 or more, tails) = 3/12 = 1/4.
   e. The favorable outcomes are: 1T, 2T, 3T, 4T, 5T, 6T. So P(any number, tails) = 6/12 = 1/2. (Since the die can show any number, the probability is just the probability of getting tails with the coin.)

5. a.

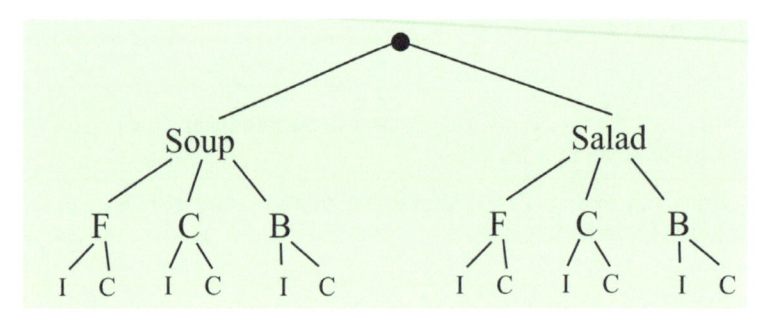

   ```
                    •
              /         \
           Soup         Salad
          / | \         / | \
         F  C  B       F  C  B
        /\ /\ /\      /\ /\ /\
        I C I C I C   I C I C I C
   ```

   Outcomes: SoFI SoFC SoCI SoCC SoBI SoBC   SaFI SaFC SaCI SaCC SaBI SaBC

   Notice that listing all the possible outcomes at the bottom of the tree diagram is completely optional.

   b. P(ice cream) = 6/12 = 1/2.

   c. P(soup, fish and cake) = 1/12.

   d. P(soup and fish) = 2/12 = 1/6.

   e. P(salad and ice cream) = 3/12 = 1/4.

   f. P(no chicken) = 8/12 = 2/3.

   g. P(no fish or ice cream) = 4/12 = 1/3.

Counting the Possibilities, cont.

6.

Second marble → First marble ↓	R	R	G	G	B
R	RR	RR	RG	RG	RB
R	RR	RR	RG	RG	RB
G	GR	GR	GG	GG	GB
G	GR	GR	GG	GG	GB
B	BR	BR	BG	BG	BB

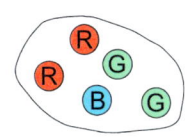

a. P(red, then green) = 4/25.
b. P(green, then red) = 4/25.
c. P(not blue, not blue) = 16/25.
d. P(not red, not red) = 9/25.

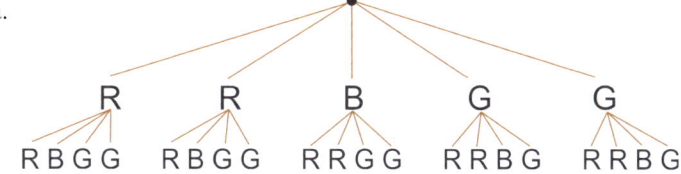

7. a.

```
            R       R       B       G       G
          RBGG    RBGG    RRGG    RRBG    RRBG
```

b. P(not red, not red) = 6/20 = 3/10.
c. P(red, then green) = 4/20 = 1/5.
d. P(green, then red) = 4/20 = 1/5.
e. P(red and green in either order) = 1/5 + 1/5 = 2/5.
f. Results will vary. Check the student's results.

8. a.

Sample space:

3,3 3,4 3,5 3,7 3,8 3,9
4,3 4,4 4,5 4,7 4,8 4,9
5,3 5,4 5,5 5,7 5,8 5,9
7,3 7,4 7,5 7,7 7,8 7,9
8,3 8,4 8,5 8,7 8,8 8,9
9,3 9,4 9,5 9,7 9,8 9,9

b. P(4; 9) = 1/36.
c. P(even; 7) = 2/36 = 1/18.
d. P(even; odd) = 8/36 = 2/9.
e. P(less than 6; more than 6) = 9/36 = 1/4.
f. P(not 6; not 6) = 1.
g. P(both digits are the same) = 6/36 = 1/6.

Counting the Possibilities, cont.

9. a.

```
                    •
       ┌────┬────┬──┴─┬────┬────┐
       B1   B2   B3   B4   G1   G2
      /|||\ /|||\ /|||\ /|||\ /|||\ /|||\
   B2 B3 B4 G1 G2  B1 B3 B4 G1 G2  B1 B2 B4 G1 G2  B1 B2 B3 G1 G2  B1 B2 B3 B4 G2  B1 B2 B3 B4 G1
```

b. P(both are girls) = 2/30 = 1/15.
c. P(both are boys) = 12/30 = 2/5.
d. P(girl, then boy) = 8/30 = 4/15.
e. P(boy, then girl) = 8/30 = 4/15.

Check: The sum of the four probabilities above is 1/15 + 2/5 + 4/15 + 4/15 = 1.

f. P(one boy, one girl) = P(girl, then boy) + P(boy, then girl) = 4/15 + 4/15 = 8/15.

10. a. The possible outcomes are: TT, TH, HT, HH.
    b. The probability of each of the four outcomes is 1/4.
    c. Predict about how many times you would expect to see each outcome: <u>50 times</u>.
    d. See two example results below:

Outcome	Frequency	% of total tosses
TT	54	27%
TH	52	26%
HT	40	20%
HH	54	27%
TOTALS	200	100%

Outcome	Frequency	% of total tosses
TT	47	23.5%
TH	44	22%
HT	53	26.5%
HH	56	28%
TOTALS	200	100%

d. Yes, the observed frequencies should be fairly close to those predicted by theory. If they weren't, perhaps the coins could be weighted (not fair) or perhaps the manner of tossing them makes them more likely to land on one side than the other, or maybe there simply weren't enough tosses. It takes several hundred tosses before the experimental probabilities get within a few percentage points of the theoretical ones.

Puzzle corner

a. P(both questions correct) = 1/16.
   There are 4 · 4 = 16 possible ways to answer the two questions. If Andy chooses randomly, each outcome is equally likely, yet only one of the outcomes represents the correct answers.

b. P(all correct) = $1/4^5$ = 1/1,024.
   Imagine making a tree diagram for this "experiment": At each node (question), it has four new branches, so the number of possibilities increases fourfold at each question. Thus there are 4^5 = 1,024 possible ways to answer the test. Of course only one of them is correct, and thus the probability of getting the correct answers by answering randomly is 1/1,024.

Using Simulations to Find Probabilities, p. 27

1. Since this is a chance process, results may vary quite a bit if the number of repetitions is fairly low (100 - 300). The example results below come from running the simulation 500 times using the spreadsheet provided with the download version of the curriculum. The student's results will vary from these more or less, depending on the number of repetitions.

Outcome	Frequency	Experimental probability
0 F 10 M	3	0.6%
1 F 9 M	5	1%
2 F 8 M	25	5%
3 F 7 M	57	11.4%
4 F 6 M	95	19%
5 F 5 M	122	24.4%
6 F 4 M	104	20.8%
7 F 3 M	65	13%
8 F 2 M	20	4%
9 F 1 M	4	0.8%
10 F 0 M	0	0%
TOTALS	500	100%

Outcome	Frequency	Experimental probability
0 F 10 M	0	0%
1 F 9 M	5	1%
2 F 8 M	22	4.4%
3 F 7 M	49	9.8%
4 F 6 M	112	22.4%
5 F 5 M	108	21.6%
6 F 4 M	111	22.2%
7 F 3 M	68	13.6%
8 F 2 M	17	3.4%
9 F 1 M	8	1.6%
10 F 0 M	0	0%
TOTALS	500	100%

Outcome	Frequency	Experimental probability
0 F 10 M	0	0%
1 F 9 M	3	0.6%
2 F 8 M	22	4.4%
3 F 7 M	64	12.8%
4 F 6 M	96	19.2%
5 F 5 M	128	25.6%
6 F 4 M	99	19.8%
7 F 3 M	58	11.6%
8 F 2 M	23	4.6%
9 F 1 M	6	1.2%
10 F 0 M	1	0.2%
TOTALS	500	100%

b. Answers will vary. If the simulation was run 100 times, you can expect this probability to fall within a range from about 16% to 30%. The theoretical probability is about 24.61%.

c. Answers will vary. If the simulation was run 100 times, you can expect this probability to fall within a range from about 15% to 27%. The theoretical probability is about 20.51%.

d. P(at least 3 females and at least 3 males) = P(3F, 7M) + P(4F, 6M) + P(5F, 5M) + P(6F, 4M) + P(7F, 3M).
Answers will vary. If the simulation was run 100 times, you can expect this answer to fall within a range from about 83% to 95%. The theoretical probability is about 89.06%.

e. P(1 or 2 of one sex, 9 or 8 of the other) = P(1F, 9M) + P(2F, 8M) + P(8F, 2M) + P(9F, 1M).
Answers will vary. If the simulation was run 100 times, you can expect this probability to fall within a range from about 5% to 15%. The theoretical probability is about 10.74%.

Using Simulations to Find Probabilities, cont.

2. The designs will vary. Check the student's design. Here are two examples:

   (1) Toss 6 coins to represent choosing 6 students. Let heads represent a student who completed homework on time and tails represent one who didn't.

   (2) Use a computer program to generate sets of 6 random digits (zeros or ones). Let 1 represent a student who completed homework on time, and let 0 represent one who didn't.

   The three example results below come from running a simulation that uses random digits zero and one 100 times.

Results of simulation		
Students who finished homework	Relative Frequency	Experimental probability
0	1/100	1%
1	6/100	6%
2	23/100	23%
3	28/100	28%
4	30/100	30%
5	9/100	9%
6	3/100	3%
TOTALS	100	100%

Results of simulation		
Students who finished homework	Relative Frequency	Experimental probability
0	1/100	1%
1	12/100	12%
2	22/100	22%
3	33/100	33%
4	19/100	19%
5	9/100	9%
6	4/100	4%
TOTALS	100	100%

Results of simulation		
Students who finished homework	Relative Frequency	Experimental probability
0	2/100	2%
1	9/100	9%
2	25/100	25%
3	37/100	37%
4	22/100	22%
5	5/100	5%
6	0/100	0%
TOTALS	100	100%

   a. Answers will vary. If the simulation was run 100 times, you can expect this probability to fall within a range from about 13% to 36%. The theoretical probability is about 23.4%.

   b. Answers will vary. If the simulation was run 100 times, you can expect this probability to fall within a range from about 4% to 16%. The theoretical probability is about 9.4%.

   c. Answers will vary. If the simulation was run 100 times, you can expect this probability to fall within a range from about 0% to 5%. The theoretical probability is about 1.6%.

   d. P(at most 2 completed homework on time) = P(0 completed it) + P(1 completed it) + P(2 completed it).
      Answers will vary. If the simulation was run 100 times, you can expect this probability to fall within a range from about 26% to 45%. The theoretical probability is about 34.38%.

   e. P(at least 3 completed homework on time) = 100% − P(at most 2 completed homework on time).
      Answers will vary. If the simulation was run 100 times, you can expect this probability to fall within a range from about 56% to 76%. The theoretical probability is about 65.63%.

Using Simulations to Find Probabilities, cont.

3. The three example results below come from running the simulation 50 times. The student's results may vary from them quite a bit since this is a chance process and the number of repetitions (50) is very low.

Students who finished homework	Relative Frequency	Experimental probability
0	0/50	0%
1	0/50	0%
2	1/50	2%
3	9/50	18%
4	17/50	34%
5	15/50	30%
6	8/50	16%
TOTALS	50	100%

Students who finished homework	Relative Frequency	Experimental probability
0	0/50	0%
1	1/50	2%
2	4/50	8%
3	8/50	16%
4	20/50	40%
5	15/50	30%
6	2/50	4%
TOTALS	50	100%

Students who finished homework	Relative Frequency	Experimental probability
0	0/50	0%
1	1/50	2%
2	7/50	14%
3	8/50	16%
4	16/50	32%
5	12/50	24%
6	6/50	12%
TOTALS	50	100%

a. Add the probabilities for 0, 1, and 2 students completing the homework on time. Answers will vary. If the simulation was run 50 times, you can expect this probability to fall within a range from about 2% to 16%. The theoretical probability is about 7.1%.

b. Answers will vary. Note that the events "at most two" and "at least three" will sum to 100%. So the easiest way to solve this is to subtract the probability that you calculated in (3a) from 100%. If the simulation was run 50 times, you can expect this probability to fall within a range from about 88% to 98%. The theoretical probability is about 92.9%.

c. Subtract the probability that all six of them have completed homework from 100%. Answers will vary. If the simulation was run 50 times, you can expect this probability to fall within a range from about 78% to 94%. The theoretical probability is about 88.2%.

d. This is the same as the probability that at least 4 of them have completed homework on time. Add the probabilities for 4, 5, and 6 students. Answers will vary. If the simulation was run 50 times, you can expect this probability to fall within a range from about 64% to 82%. The theoretical probability is about 74.4%.

4. Below are three example results from running the simulation 200 times. The student's results may vary from them since this is a chance process and the number of repetitions (200) is fairly low. For a comparison, the theoretical probabilities are also given.

Students who finished homework	Relative Frequency	Experimental probability
0	0/200	0%
1	4/200	2%
2	14/200	7%
3	50/200	25%
4	57/200	28.5%
5	57/200	28.5%
6	18/200	9%
TOTALS	200	100%

Students who finished homework	Relative Frequency	Experimental probability
0	1/200	0.5%
1	4/200	2%
2	10/200	5%
3	35/200	17.5%
4	57/200	28.5%
5	68/200	34%
6	25/200	12.5%
TOTALS	200	100%

Students who finished homework	Relative Frequency	Experimental probability	Theoretical probability
0	0/200	0%	0.1%
1	3/200	1.5%	1.2%
2	14/200	7%	6.0%
3	30/200	15%	18.5%
4	65/200	32.5%	32.4%
5	62/200	31%	30.3%
6	26/200	13%	11.8%
TOTALS	200	100%	100%

a. This is the same as the probability that at least 3 of them have completed homework on time. Add the probabilities for 3, 4, 5, and 6 students. Expect a result from 89% to 95%. The theoretical probability is 92.95%.

b. Add the probabilities for 4, 5, and 6 students. Expect a result from 67% to 80%. The theoretical probability is 74.43%.

47

Using Simulations to Find Probabilities, cont.

5. a. Designs will vary. Check the student's design. For example: "To represent the twenty people who typically come to the clinic that day, generate a sequence of 20 random digits from 0 to 9. Let the first digit be the first person, the second digit the second person, and so on. The probability of type A blood is 40%, so if the digit is 0, 1, 2, or 3, then that person has type A blood, if the digit is from 4 to 9, then the person has some other blood type."

b. Results will vary. Check the student's results. Here are three example results:

Results of simulation		
Event	Frequency	Experi-mental probability
A _ _ _	35	35%
XA _ _	30	30%
XXA _	16	16%
XXXA	7	7%
XXXX	12	12%
TOTALS	100	100%

Results of simulation		
Event	Frequency	Experi-mental probability
A _ _ _	39	39%
XA _ _	22	22%
XXA _	11	11%
XXXA	14	14%
XXXX	14	14%
TOTALS	100	100%

Results of simulation		
Event	Frequency	Experi-mental probability
A _ _ _	41	41%
XA _ _	23	23%
XXA _	14	14%
XXXA	7	7%
XXXX	15	15%
TOTALS	100	100%

c. To find the probability that it will take 1, 2, or 3 donors until you find one with blood type A, add the probabilities for A _ _ _, XA _ _, and XXA _. Expect a result from about 69% to 88%. The theoretical probability is 78.4%.

d. The probability that it will take exactly 4 donors until you find one with blood type A is the outcome XXXA. Expect a result from about 3% to 14%. The theoretical probability is 8.64%.

e. The probability that it will take more than 4 donors to find one with blood type A is the outcome XXXX. Expect a result from about 7% to 21%. The theoretical probability is 12.96%.

f. The probability that it will take at least 4 donors to find one with blood type A is the sum of the probabilities for outcomes XXXA (it takes exactly 4 donors) and XXXX (it takes more than 4 donors). Expect a result from about 10% to 28%. The theoretical probability is 21.6%.

Using Simulations to Find Probabilities, cont.

6. a. Designs will vary. Check the student's design. For example: "Generate sequences of three random numbers where the random numbers have values from 1 to 4: 1 = strawberry, 2 = lemon, 3 = blackberry, and 4 = apple." This is how you can set up the random number generator at https://www.random.org/integers/ to do that:

   > **Part 1: The Integers**
   >
   > Generate 300 random integers (maximum 10,000).
   >
   > Each integer should have a value between 1 and 4 (both inclusive; limits ±1,000,000,000).
   >
   > Format in 3 column(s).

   b. Results will vary. Check the student's results. The image below shows sorting in Excel, which might be somewhat helpful. You will still need to look through the outcomes carefully to find the favorable ones for each exercise.

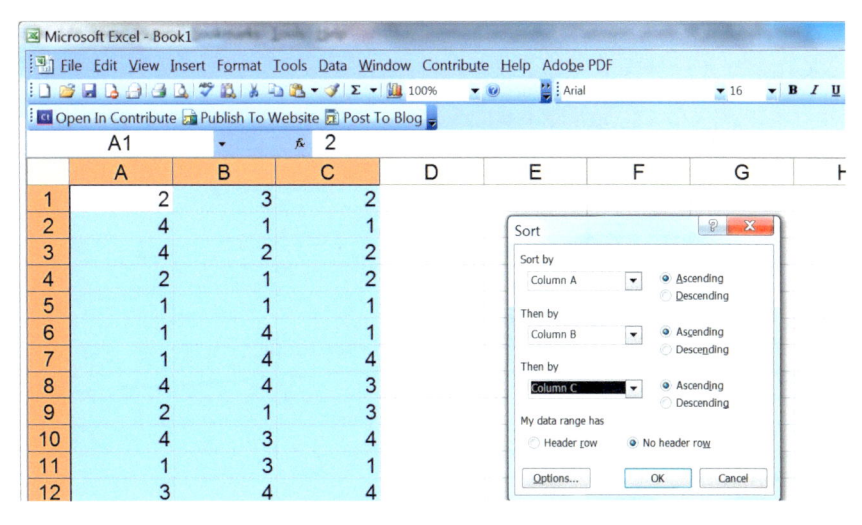

   c. P(none are strawberry). The theoretical probability is $0.75 \times 0.75 \times 0.75 = 42.1875\%$.
      If you run the simulation 100 times, expect a result from about 38% to 51%.
      If you run the simulation 500 times, expect a result from about 39% to 45%.

   d. P(exactly 2 are strawberry). The theoretical probability is $0.25 \times 0.25 \times 0.75 + 0.25 \times 0.75 \times 0.25 + 0.75 \times 0.25 \times 0.25 = 14.0625\%$.
      If you run the simulation 100 times, expect a result in a range from about 5% to 19%.
      If you run the simulation 500 times, expect a result in a range from about 12% to 16%.

   e. P(all 3 are strawberry): The theoretical probability is $0.25 \times 0.25 \times 0.25 = 1.5625\%$.
      If you run the simulation 100 times, expect a result in a range from about 0% to 4%.
      If you run the simulation 500 times, expect a result in a range from about 0.6% to 3%.

   f. P(none are lemon or blackberry): The theoretical probability is $0.5 \times 0.5 \times 0.5 = 12.5\%$.
      If you run the simulation 100 times, expect a result in a range from about 8% to 22%.
      If you run the simulation 500 times, expect a result in a range from about 9% to 16%.

   g. P(all three are the same flavor): The theoretical probability is: $(0.25 \times 0.25 \times 0.25)_{\text{Strawberry}} + (0.25 \times 0.25 \times 0.25)_{\text{Lemon}} + (0.25 \times 0.25 \times 0.25)_{\text{Blackberry}} + (0.25 \times 0.25 \times 0.25)_{\text{Apple}} = 6.25\%$.

      If you run the simulation 100 times, expect a result in a range from about 2% to 14%.
      If you run the simulation 500 times, expect a result in a range from about 4% to 8%.

Probabilities of Compound Events, p. 33

1. a. P(THT) = (1/2) · (1/2) · (1/2) = 1/8.
   b. P(X, H, X) = 1/2, which is just the probability of getting heads on any one toss.
   c. There are three favorable outcomes: THH, HTH, HHT, so the probability is 3/8.

2. a. P(red, then green) = (2/9) · (5/9) = 10/81.
   b. P(green, then red) = (5/9) · (2/9) = 10/81.
   c. P(not blue, not blue) = (7/9) · (7/9) = 49/81.
   d. P(not green, not green) = (4/9) · (4/9) = 16/81.

3. a. P(not pink for five days) = $(7/8)^5 \approx 0.5129 \approx 51\%$.
   b. P(not pink for ten days) = $(7/8)^{10} \approx 0.2631 \approx 26\%$.

4. a. 4 · 4 = 16 different numbers.
   b. P(35) = 1/16.
   c. The favorable outcomes are: 36, 45, 54, and 63, so the probability is 4/16 = __1/4__.

5. a. P(go to park and not rain) = (1/2) · (4/5) = 4/10 = 2/5.
   b. P(go to park and rain) = (1/2) · (1/5) = 1/10.

6. At first, there are four cards with an even number and eight cards in total to choose from. So P(even) = 4/8 = __1/2__. After one card with an even number has been drawn, there are seven cards left, and one of them is 7. So P(7) is __1/7__. Then we multiply the two probabilities to get the probability of both events: P(even, 7) = (1/2) · (1/7) = __1/14__.

7. a. P(heart, heart) = (3/9) · (2/8) = (1/3) · (1/4) = 1/12.
   b. P(star, cross) = (2/9) · (4/8) = (2/9) · (1/2) = 1/9.
   c. P(not heart, not heart) = (6/9) · (5/8) = (2/3) · (5/8) = 5/12.
   d. P(star, not star) = (2/9) · (7/8) = 14/72 = 7/36.

8. a. P(both green) = (5/9) · (4/8) = (5/9) · (1/2) = 5/18.
   b. P(neither is green) = (4/9) · (3/8) = 12/72 = 1/6.
   c. P(one is green) = 1 − P(both green) − P(neither is green) = 1 − 5/18 − 1/6 = 1 − 5/18 − 3/18 = 10/18 = 5/9.

9. a. P(white, white) = (10/24) · (9/23) = (5/12) · (9/23) = (5/4) · (3/23) = __15/92__ ≈ 16.3%.
   b. P(black, black) = (14/24) · (13/23) = (7/12) · (13/23) = __91/276__ ≈ 33.0%.
   c. P(black, white) = (14/24) · (10/23) = (7/12) · (10/23) = (7/6) · (5/23) = __35/138__ ≈ 25.4%.
   d. P(white, black) = (10/24) · (14/23) = __35/138__ ≈ 25.4%.

   CHECK. The four probabilities above *do* total 1 or 100%.

   e. P(matching socks) = P(white, white) + P(black, black) = 15/92 + 91/276 = (45 + 91)/276 = 136/276 = __34/69__ ≈ 49.3%.
   f. P(not matching socks) = 1 − P(matching socks) = 1 − 136/276 = 140/276 = __35/69__ ≈ 50.7%.

10. a. P(AAA) = (4/52) · (3/51) · (2/50) = (1/13) · (1/17) · (1/25) = 1/5525 ≈ 0.000181 = 0.0181%.
    b. P(AXX) = (4/52) · (48/51) · (47/50) = 376 / 5525 ≈ 6.8%.
    c. P(XAX) = (48/52) · (4/51) · (47/50) = 376 / 5525 ≈ 6.8%.
    d. P(XXA) = (48/52) · (47/51) · (4/50) = 376 / 5525 ≈ 6.8%.
    e. P(exactly one ace) = 3 · 376 / 5525 = 1128 / 5525 ≈ 20.4%.

Puzzle corner. P(matching pair) = P(white, white) + P(brown, brown) + P(black, black)
= (8/27) · (7/26) + (9/27) · (8/26) + (10/27) · (9/26) = (56 + 72 + 90)/702 = 218/702 ≈ 31.05%.

Review, p. 37

1. a. P(not math, science, or English) = P(social studies, art, or music) = 9/25.
   b. P(math) = 7/25.
   c. There are 13 boys, and four of them chose math. So P(a boy's favorite subject is math) = 4/13.
   d. There are 12 girls, and three of them chose math. So P(a girl's favorite subject is math) = 3/12 = 1/4.

2. You can find these probabilities by listing and counting the favorable outcomes, or since these events are compound events, you can also find them by multiplying the probabilities of the individual events.

   a. P(5, 6) = (1/6) · (1/6) = 1/36.

   b. There are 9 favorable outcomes: (2, 2), (2, 4), (2, 6), (4, 2), (4, 4), (4, 6), (6, 2), (6, 4), and (6, 6), so P(even, even) = 9/36 = 1/4.

      Or: P(even, even) = (1/2) · (1/2) = 1/4.

   c. There are 4 favorable outcomes: (5, 5), (5, 6), (6, 5), and (6, 6), so P(at least 5, at least 5) = 4/36 = 1/9.

      Or: P(at least 5, at least 5) = (2/6) · (2/6) = 4/36 = 1/9.

   d. There are 6 favorable outcomes: (1, 1), (1, 2), (2, 1), (2, 2), (3, 1), and (3, 2), so P(at most 3, at most 2) = 6/36 = 1/6.

      Or: P(at most 3, at most 2) = (3/6) · (2/6) = (1/2) · (1/3) = 1/6.

3. a. P(1) = 8/60 = 2/15. P(4) = 13/60.

   b. Because rolling a die is a chance or random process: you never know what you will get when you roll it. Rolling a die six times does not guarantee that you get one of each number. As the number of repetitions increases, the relative frequencies (experimental probabilities) do get closer and closer to the theoretical probabilities of 1/6. However, 60 is not a large number of repetitions, so we expect the experimental probabilities to vary a lot from 1/6.

4. a.

Outcome	Frequency	Experimental Probability (%)	Theoretical Probability (%)
HH	38	19%	25%
HT	53	26.5%	25%
TH	46	23%	25%
TT	63	31.5%	25%
TOTALS	200	100%	100%

   b. The experimental probabilities would be much closer to the theoretical ones (much closer to 25%).

5. You can find these probabilities by listing and counting the favorable outcomes or by multiplying the probabilities of the individual events. There are a total of 7 · 6 = 42 possible outcomes, each being equally likely.

   a. There is only one favorable outcome (purple, orange), so the probability is 1/42.
      Or: P(purple, orange) = (1/7) · (1/6) = 1/42.

   b. There are five favorable outcomes: (red, blue), (red, purple), (red, orange), (red, yellow), and (red, mint), so the probability is 5/42.
      Or: P(red, not pink) = (1/7) · (5/6) = 5/42.

   c. There are 30 favorable outcomes: each of the six colors *red, blue, purple, pink, orange,* and *yellow* combined with five of those colors so that the same color is not chosen twice. Thus the probability is 30/42 = 5/7.
      Or: P(not mint, not mint) = (6/7) · (5/6) = 5/7.

Probability, Grade 7
Alignment to the Common Core Standards

The table below lists each lesson and next to it the relevant Common Core Standard.

Lesson	Page number	Standards
Probability	13	7.SP.5 7.SP.7
Probability Problems from Statistics	16	7.SP.7
Experimental Probability	18	7.SP.6 7.SP.7
Counting the Possibilities	21	7.SP.7 7.SP.8
Using Simulations to Find Probabilities	27	7.SP.8
Probabilities of Compound Events	33	7.SP.8
Review	37	7.SP.6 7.SP.7 7.SP.8

Made in United States
Troutdale, OR
12/19/2023